Joseph Kendall Freitag

The Fireproofing of Steel Buildings

First Edition

Joseph Kendall Freitag

The Fireproofing of Steel Buildings
First Edition

ISBN/EAN: 9783337249229

Printed in Europe, USA, Canada, Australia, Japan

Cover: Foto ©berggeist007 / pixelio.de

More available books at **www.hansebooks.com**

THE

FIREPROOFING

OF

STEEL BUILDINGS.

BY

J. K. FREITAG, C.E.,

Author of "Architectural Engineering."

FIRST EDITION.

FIRST THOUSAND.

NEW YORK:

JOHN WILEY & SONS.

LONDON: CHAPMAN & HALL, LIMITED.

1899.

PREFACE.

THE want of any systematized and collected form of information on the subject of the development of the fireproofing of steel buildings and its present most approved and efficient methods of treatment, as recommended and used in the best practice of the day, has induced the author to offer this volume. It is hoped that it may rouse more general interest in a phase of building construction, the full importance of which, from an economic standpoint and from its close relation to the safety of human life, still largely remains to be realized in a practical and adequate way.

The subject of fireproofing is still in the interesting stage of experimentation, and there is vast opportunity for practical ingenuity to make improvement and new suggestion. Indeed the term fireproof in relation to present imperfect conditions and environments is almost utopian. A building can, at best, be only relatively fire- and water-proof. No structure can be considered as a unit, exclusive of its neighbor. But it may be assumed that a building designed and erected in accordance with the recommendations made in the following chapters, will be as nearly fireproof as the surrounding buildings and the character of the contents will permit.

It may be argued that some of the suggestions presented are not applicable to modern methods; but theoretical advantages are none the less true even though present conditions

iii

of development, in some cases, prohibit their practical application. But even if the distinctly practicable recommendations alone are followed, a structure capable of resisting severe attacks by fire and water, with a minimum of repair reconstruction after fire, will surely result. The efficiency of modern methods has been largely demonstrated by past experience, and this experience has proved that still more perfect fireproofing may result if the structure is scientifically designed and conscientiously built and equipped.

For the purpose of tracing through the important elements of fireproof design and construction, the following chapters have been arranged to cover:

Introductory and development, Chapters I and II.

Fires and tests, Chapters III and IV.

Materials, Chapters V and VI.

Planning, Chapter VII.

Details, Chapters VIII to XIII inclusive.

Equipment, Chapter XIV.

JOSEPH KENDALL FREITAG.

BOSTON, May, 1899.

CONTENTS.

CHAPTER I.

PAGE

FIRE PROTECTION.. 1

CHAPTER II.

DEVELOPMENT OF FIREPROOFING METHODS AND STEEL BUILDING CON-
STRUCTION.. 8

CHAPTER III.

FIRES IN FIREPROOF BUILDINGS.. 24

CHAPTER IV.

TESTS OF FIREPROOF FLOORS.. 49

CHAPTER V.

MATERIALS USED IN FIRE-RESISTING CONSTRUCTION............................... 76

CHAPTER VI.

PERMANENCY—CORROSION... 108

CHAPTER VII.

FIRE-RESISTING DESIGN.. 121

CHAPTER VIII.

TERRA-COTTA FLOORS—GIRDER PROTECTION....................................... 147

CHAPTER IX.

CONCRETE AND COMPOSITION FLOORS.. 178

v

CHAPTER X.

COLUMN FIREPROOFING 217

CHAPTER XI.

FIRE-RESISTING PARTITIONS... 236

CHAPTER XII.

EXTERIOR WALLS, SPANDRELS, AND WALL-COLUMNS 255

CHAPTER XIII.

ROOFS, SUSPENDED CEILINGS, FURRING 273

CHAPTER XIV.

EQUIPMENT FOR FIRE-RESISTANCE... 292

THE FIREPROOFING OF STEEL BUILDINGS.

CHAPTER I.

FIRE PROTECTION.

INASMUCH as the annual fire losses in the United States are steadily increasing with the development and population of the country, this continued waste of resources must constitute one of the great fields for improvement that ought to claim the attention of serious thought at the present time. Most of our cities have passed through the stage of early and temporary development, and attention should now be turned to the permanent and fire-resisting protection of property which has hitherto been improved according to the inefficient standard of non-fireproof eras.

For the ten years from 1888 to 1897 inclusive, the annual fire loss in the United States has averaged $132,296,043, the maximum loss for any one year being $167,544,370, in 1893. The total fire loss for the year 1897 was $116,354,570. Adding to this the amount estimated as necessary to maintain protection against fire loss, by means of the fire departments, protective departments, and insurance companies, the total annual outlay occasioned by fire will approximate $170,000,000.

Considering the fire-insurance companies, it is found that the losses paid by these associations as their share of the annual fire waste will average $80,763,375 yearly for the same ten years. The greatest losses paid in any one year were in

1893, when $105,994,577 were paid out in reimbursements for fire insurance. For the year 1897 the insurance losses were $66,722,140.

It is manifestly impossible to assign any definite proportion of the losses by fire to the careless design or execution of details of construction. Certain unavoidable adverse conditions will always enter into the problem of a perfect fireproof building in order to render it practicable from an economic standpoint, even if it is unsatisfactory to the theorist. But a very great number of causes of fire, or serious damage by fire, are comparatively easy of correction.

It has been stated by underwriters that fully forty per cent. of all fires can be directly traced to causes which might easily be prevented. Defective flues, for instance, which constitute one of the most ordinary causes, might be easily eliminated by more careful construction. But whatever the cause, careful building construction will do much to modify the great losses which so frequently occur. Imperfect construction and careless fireproofing make a building unfit to assume its share of the responsibility in checking a fire of however trivial a beginning. The ravage is passed on from structure to structure, whereas one thoroughly incombustible building might have formed an efficient barrier to stop the devastation.

The burning of the Hotel Windsor in New York City on March 17, 1899, furnishes one of the most striking object-lessons in many years of the tremendous risk to life and property which is occasioned by ordinary building construction. Although used as a place of abode by several hundreds of persons, the structure was not fireproof, nor were even elementary precautions taken, such as placing fire-stops in the hollow walls and partitions. Wooden beams, studding and lathing, with hollow spaces in every floor and partition, constituted a construction than which none better could be devised for the rapid and fatal spread of fire. And the statement is perhaps none too strong that ninety-nine per cent. of the buildings erected throughout this country are "as perfectly adapted to facilitate the destruction of the buildings by fire as if they had been designed with that end in view."

A great contrast is provided by the methods employed in European countries. The greater number of fires and the higher percentage of values destroyed by fire in the United States over Europe is easily accounted for by the better construction which prevails in European practice. The reports of U. S. consuls on fire and building regulations in foreign countries issued in 1892 by the State Department at Washington, in reply to questions propounded by the Government at the instance of the National Board Committee on Statistics, leave no doubt as to the superiority and efficiency of the average European construction.

"In most instances the fires in leading European cities are confined to the floor on which they originated, and it is rarely that they pass beyond the building. Thus we find it stated that in Vienna in 1889 all of the fires (570) were confined to the floor; that in London in 1889 there were 2,892 fires, all but 9 of which were confined to the building. In Hamburg in 1890 only 10 out of 682 fires went beyond the building, and in Glasgow all but 14 in a total of 504 were confined to the floor. In Leeds, Sheffield, Bristol, and Newcastle, England, every fire was confined to the building, and this was also the case at Rome, Florence, Turin, Copenhagen, Brussels, Trieste, Antwerp, and Havre." *

Our fire losses cannot be met by the simple provision of insurance. Insurance is only a means whereby the community bears the loss of the individual. It is a palliative for the evil, not a cure. Nor is it possible to effect a change simply by increasing the rates of insurance. The insured must cooperate

---

* "Proceedings of the 27th Annual Meeting of the National Board of Fire Underwriters."

with the underwriters in bringing about a better class of building construction, and this will obtain as soon as the individual
realizes what a proportion of a reckless waste he is made to
bear.

Up to the present time fireproofing methods have very
generally been confined to buildings of a public or semi-public
nature, such as government buildings and a few theatres and
places of public amusement or gathering, or to factories, stores,
warehouses, and office buildings. The latter class of structures,
devoted to mercantile or office purposes, has received by far
the greatest attention from a fireproofing standpoint. The
fireproofing of the modern office building has been largely a
matter of necessity, occasioned by the great height which
has been demanded to secure an adequate financial return
upon the high-priced property. But in store and factory
buildings and warehouses the recourse to fireproof methods
has been brought about by the additional security to life,
property, and business, the more permanent construction, and
the decreased insurance rates which attend the use of fire-
resisting construction. In office buildings in the city of Boston many of the older non-fireproof buildings have insurance
rates of from 28 cents to 47 cents per $100, on the 80 per cent.
insurance-guarantee basis, while the more modern, distinctly
fire-resisting office buildings are rated as low as 14 cents to 20
cents per $100. For warehouses the insurance rating on fire-
proof buildings will average from 14 cents to 16 cents.

In speaking of office and commercial buildings, Mr. W. L.
B. Jenney, in an address before the Fire Underwriters' Association of the Northwest, called attention to the duties of the
underwriters in the matter of fireproofing, and suggested that
they should vary the rate of insurance for these classes of
buildings according to the character of the fireproofing
used. Competition between the fireproofing companies has
been so severe as to reduce the price of their output, in very

many instances at the expense of quality. He advocated that the insurance companies prepare standard specifications governing the character of the fireproofing, and the construction and putting on of these materials. If owners of buildings would not then follow these specifications, under careful inspection, they should be subject to higher rates.

While this would not wholly remedy much of the present apathy towards fireproof construction, benefits would certainly result from a careful following of these suggestions. In certain respects the insurance companies have great advantages, from their experience gained through losses, in estimating the value of different kinds of materials and methods. They could largely enforce their views from the fact that investors would give heed if higher insurance rates followed a neglect of the requirements demanded.

The relative cost of fireproof and non-fireproof buildings will not be discussed here. The questions of cost of real estate, building construction, insurance rates,. and many other local or special conditions will greatly influence the relative economy in each particular case. The present low prices of building materials, the cheapness and facility for obtaining structural steel and terra-cotta, or composition or concrete constructions, combined with the ever-present tendency to lessen the cost of all improvements, will do much to cause fireproofing. to be gradually applied to all classes of structures, public or private, whether of low cost or expensive construction. Several notable cases of residence fires in New York City, attended by large loss of life, have begun to awaken the owners of residences to the necessity of making private dwellings fireproof. As an immediate result of these fires, the plans for several residences which, at the time, were designed as non-fireproof buildings, have since been changed to thorough fireproof construction.

The century about to open must show a more marked

improvement in the field of fireproofing than in any other detail of building improvement. The systematic study which is being applied to this field is illustrated by the work undertaken by "The British Fire Prevention Committee." This consists of an association of English architects and engineers, and men prominent in the fields of art, science, and mercantile affairs, who have undertaken the collection and publication of studies and papers bearing upon fireproof design and construction. There have already been issued twelve papers from the pens of both English and American authorities, and many other descriptive articles are in preparation. The publications must be of the utmost value to all who are interested in the development of this subject.

In the following chapters the author will endeavor to show what has so far been accomplished in the fireproofing of steel buildings. Steel skeleton construction and the development of its systematic fireproofing will be traced from earlier crude attempts down to the methods now in use for protecting the structural framework in steel buildings of fire-resisting design. Attention will also be called to such reliable and scientific tests as have been made on floor-, column-, and partition-construction, and the lessons to be learned from notable fires in buildings designed to resist fire- and water-attacks will be brought to notice.

Many important questions relative to fire-resisting construction undoubtedly still remain open, and the answers will only be found in such fires in "fireproof" buildings as those which have occurred in Chicago, Pittsburg, and New York City. But from the past examples some reasonably final conclusions may be drawn in reference to a number of features in the general problem of fireproofing.

Fireproofing has hitherto largely been a matter of theory. Modern systems have been subjected to the tests of but few large fires. Many tests have been made by manufacturers.

and some by disinterested parties, and such of these as are valuable will be mentioned. Most of these, however, have been on a small scale, and while probably giving valuable data, they have fallen far short in value as compared to such tests as were afforded by the Home Life Insurance office building in New York, or the Horne buildings in Pittsburg.

The term "fireproof" is but relative. If buildings cannot be constructed in which a destructive fire, originating within, is practically impossible, or which will withstand a severe fire in the immediately adjacent structures, then it becomes of the greatest importance to ascertain how near that end can be attained. There is little inducement to the owner to employ fireproof construction if it means loss on the fireproof building, and safety beyond it.

Destruction by fire, either internal or external, may be mitigated or prevented by the best possible general design, by careful attention to the matter of details in providing the most reliable protection that can be given by fire-resisting materials properly applied, by preventing the very high temperatures which might otherwise be developed through fire by admitting only the very smallest quantities of combustible materials, and by proper equipment to care for incipient fires or to aid the efforts of the fire department when called upon to cope with severe internal or external fire.

CHAPTER II.

DEVELOPMENT OF FIREPROOFING METHODS AND STEEL BUILDING CONSTRUCTION.

THE earliest buildings in this country of a fireproof nature were probably those used for public purposes, for the construction of which ample funds were usually provided, allowing the use of structures covering considerable area but comparatively moderate in height. In these buildings the walls were made of sufficient thickness and stability to allow the use of vaulted construction in the design of the floors. Solid brick arches were generally employed, giving rise to great extravagance in space and dead weight. To obtain the requisite strength many courses of brick were required, and long spans necessitated a great depth of flooring. A good example of such construction was to be seen in the buildings adjoining Independence Hall, Philadelphia. These buildings, known as "State House Row," were erected in the year 1813, and when they were recently torn down to allow replacement, it was found that the brick arches which had been employed in the floor construction averaged in weight about 217 pounds per square foot.

But even these structures, built with the objects of permanency and solidity, were generally covered with a roof of wooden construction, showing that any fireproofing qualities which may have existed were not the result of a well-developed plan, aiming to secure a thoroughly fireproof building. It is only in comparatively recent years that fire-resisting con-

8

struction has begun to attain to the dignity of being distinguished as a specialty.

Origin of Present Methods.—The earliest application of the present general system of fire-resisting construction was the use of the brick arch sprung between iron beams. The first known example of this construction was in the cotton-mill of Messrs. Phillips & Lee of Manchester, England, erected in 1801.* The beams were cast-iron, spaced about 10 feet on centres.

Brick Arches.—On the invention of the iron I-beam in France and England in 1853, the manufacture of this product was soon introduced into this country. The first iron I-beams were rolled at Trenton, N. J., in 1854, while the first steel beams were rolled by the Carnegie Steel Co. as late as 1885.

The earliest attempts at systematic fireproofing were through the use of wrought-iron beams and brick arches. The employment of wrought-iron I-beams permitted a lessening of the distance between the walls to any desirable span, so that the former deep and heavy brick arches could now be made thinner, and consequently lighter. The depth was finally reduced to one course of brick, or one "rowlock." This brought the weight down to about 33 lbs. or 35 lbs. per square foot when set, to which had to be added the concrete filling, varying in depth, but averaging 8 to 10 lbs. for each inch in thickness (see Fig. 1). The total weight per square foot was seldom less than 75 lbs.

This was a great advance over the thicker solid brick arches, or over the use of cast-iron beams, but this construction was still too heavy, and open to other decided objections, particularly in the exposure of the lower flanges of the beams.

---

* "On the Application of Cast- and Wrought-iron to Building Purposes," by W. Fairbairn. London, 1854.

The framework for the interiors of such buildings was made of cast-iron columns, iron girders, and the iron floor-beams which carried the brick arches. This left the lower flanges of the beams entirely exposed to the possible action of fire. The cast columns, too, were generally unprotected. In

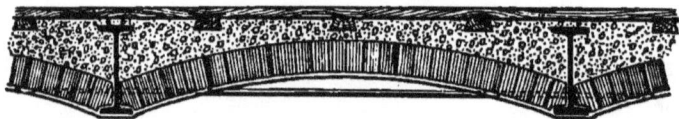

FIG. 1.—Brick Arch.

case of fire, the expansion of the iron beams was usually sufficient to damage seriously or destroy entirely the surrounding masonry walls. This would not ordinarily be the case with good walls in timber construction.

Corrugated-iron Arches.—An attempt to reduce the cost and dead weight in the brick-arch method led to the introduction of the arch in which sheets of corrugated-iron, bent to a radius, were sprung between the lower flanges of the I-beams. This did not result in any material reduction in weight, as it simply meant the substitution of the corrugated-iron sheets with more concrete for the brick arch previously used.

The use of corrugated-iron was evidently due to the early and mistaken notion that iron, in itself, was fireproof, because incombustible; but this type was still more objectionable than the previous one, in that the arches themselves were now subject to destruction by fire, as well as the beams and cast columns. Examples of this style of arch may still be seen in buildings of that period which have survived to the present time.

Introduction of Terra-cotta.—The present systems of fireproofing practically arose as a result of the great Chicago fire in 1871. Flat hollow tile arches were first patented and introduced in Chicago in 1872 by Mr. Geo. H. Johnson, and the first structure in which the Johnson arches were used was the Kendall Building, afterward called the Equitable Building. The type of floor-arch used was as shown in Fig. 2. A similar arch was also used in the Criminal Court building in Chicago.

FIG. 2.—Terra-cotta Arch in Kendall (Equitable) Building, Chicago.

At about the same time that Mr. Johnson introduced his method, a similar but heavier construction was used in New York City by Mr. Leonard F. Beckwith, in the corridors of the New York Post-office. This heavier flat arch was also used in the Coal and Iron Exchange Building on Cortlandt Street, New York City, built about 1872, which was the first extended use of the system in the East.

These arches were naturally very crude in materials and workmanship, but their introduction aroused considerable interest in building and architectural circles. The arches proved substantial, and served well the purpose of a light and fire-resisting floor. The material used was hard-burned terra-cotta.

In 1876 flat tile arches were used in the Chicago Court-house, but no progress was shown in the design or the material employed.

Improvements on Original Design.—In about the year 1878–79 a better and lighter form of hollow arch was used in the Chicago City Hall building. These blocks were made by Johnson & Co., later known as the Pioneer Fire-proof Construction Co., of Chicago. The arch was flat on the bottom and curved at the top, with the haunches levelled up with concrete, as shown in Fig. 3. This arch weighed 40 lbs. per square foot.

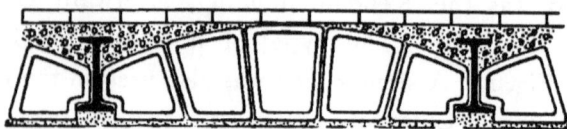

FIG. 3.—Terra-cotta Arch in Chicago City Hall.

In 1881 the Montauk Block was built in Chicago, the first of the modern office buildings erected in that city. Hollow-tile floor-arches were employed in this building in spans varying from three to four feet. The blocks were made of pure Ohio fire-clay, and were 6 ins. in depth, weighing 25 lbs. per superficial foot. See Fig. 4. The work was done by the Wight Fireproofing Co.

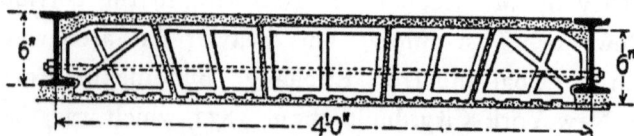

FIG. 4.—Terra-cotta Arch in Montauk Block, Chicago.

It will be noticed that all of the arches above described were so made that the bottoms of the tile blocks projected about one-half inch below the bottoms of the beams. Recessed spaces were thus left under the beam-flanges which were filled with an extra thickness of plaster at the time of applying the first coat. These surfaces under the beams, being colder than the surfaces under the hollow tile, condensed the moisture in the atmosphere along these lines, and caused soot or smoke to accumulate, so that the locations of the beams were soon indicated by black lines on the ceilings. This trouble suggested the use of protection-tiles for the beam-flanges.

First Use of Interior Webs.—Previous to the year 1883, all of the tile arches used in Chicago had been made of tiles without interior webs, excepting in the skewbacks. It was found that such tiles were not strong enough to sustain severe loads, nor the sudden strains caused by moving safes, nor even the heavy weights put upon the floors during the course of erection.

In 1883, the contract for the floors in the new building for the Mutual Life Insurance Co. of New York, on Nassau Street, New York City, was awarded to a Chicago fireproofing concern. This award constituted the first recognition of the superiority of the Western methods. A 9-in. flat arch was used throughout, each tile having one vertical and one horizontal web, the thickness of the fire-clay when burned being only half an inch. The arch is shown in Fig. 5. The arches weighed 33 lbs. per superficial foot.

FIG. 5.—Terra-cotta Arch in Mutual Life Insurance Building, New York.

This type of arch was also used in the Home Insurance Co.'s building, Chicago, 1883–84, which was the first " skeleton construction " building erected.

First Flange-protection for Beams.—The arches used in the Mutual Life Insurance Co.'s building in New York, above described as first employing interior ribs in the tiles, were also the first in which soffit tiles were used for the protection of the beam-flanges. The endeavor to prevent the discoloration by smoke, which was especially noticeable in cities where soft coal was used for fuel, led to a decided improvement in the fire-resisting qualities of the arch.

The form of flange-protection tile employed was one in which the tile strip touched the beam-flange at its outer edges only, the centre being recessed so as to leave an air-space of about one-quarter of an inch under and along the beam-flange.

Other Early Forms.—Fig. 6 shows an arch which was used in spans up to 7 ft. These blocks were 10 ins. deep,

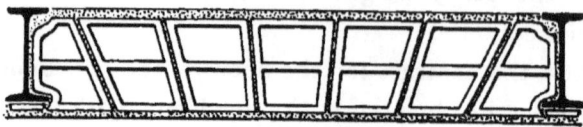

FIG. 6.—Terra-cotta Arch—Early Form.

the weight being 40 lbs. per square foot.

The arches used in the Caxton and Pontiac buildings, Chicago, are shown in Fig. 7. These blocks were made 10 ins.

FIG. 7.—Terra-cotta Arch in Caxton and Pontiac Buildings, Chicago.

deep, set between 12-in. beams. Both of these buildings were designed for printing and light manufacturing purposes.

Fig. 8 shows the arches used in the Board of Trade and Telephone buildings, Chicago. Twelve-inch blocks were set

FIG. 8.—Terra-cotta Arch in Board of Trade and Telephone Buildings, Chicago.

between 15-in. beams, 4 ins. of concrete being placed over the arches.

Fig. 9 shows the floors employed in the Indiana State Capitol building at Indianapolis, Ind. In this case the arches were 12 ins. deep, set between 12-in. beams, with only enough concrete filling to level up to the tops of the I-beams, to form a ground for the marble floors.

Introduction of Porous Tile.—Porous terra-cotta, in the form of roof-blocks and slabs for column-protection, was used as early as 1873. Porous roof-blocks set between T-irons

FIG. 9.—Terra-cotta Arch in Indiana State Capitol, Indianapolis, Ind.

were used at about this date in the Chicago Water Works building, and as column gores in the Chicago Club House.

Porous floor-blocks were first employed in the old south wing of the Patent Office, at Washington, D. C.

This product was first used on a large scale in the Chicago Auditorium. The arches were similar in section to those of hard-burned fire-clay, but the webs of the tiles were about double the thickness of those employed in the hard-burned blocks. The weight per square foot remained about the same.

Segmental Tile Arches.—Fig. 10 shows the type of floor used in the Minnesota Hospital for the Insane at Richmond,

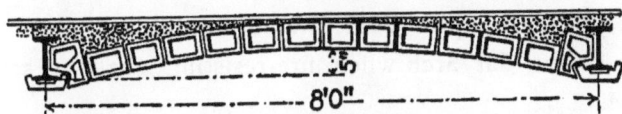

FIG. 10.—Segmental Terra-cotta Arch. Minnesota Hospital for the Insane.

Minn. The illustration shows an arch of 8-ft. span with a 5-in. rise. Spans were employed up to 14 feet.

Another early form of segmental tile arch is shown in Fig. 11. The arch blocks were 6 ins. deep with a 9-in. rise, and were used between 12-in. and 15-in. beams. The haunches were filled in with a second ring of similar blocks. Spans were employed up to 15 feet.

The first office building in which long-span segmental

arches were used was the Commerce Office Building in Chicago, where 16-ft. spans with a 9-in. rise were constructed. No beam-flange covering was used, but the ceilings were

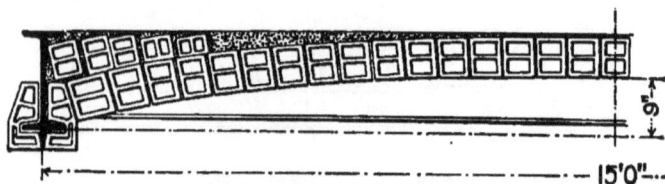

FIG. 11.—Segmental Terra-cotta Arch—Early Form.

furred down level by means of light suspended tee-irons, spaced about 12 ins. centres, on which were carried split tiles with grooved edges. The tests for strength showed satisfactory results, but the use of a double system for such long spans, to produce a level ceiling, proved very expensive.

Long-span segmental arches have continued in use in buildings where flat ceilings are not considered necessary; and where the tie-rods are spaced evenly, they may be encased in tile so as to produce a panelled effect after plastering. Segmental tile arches, however, have never superseded the flat-arch type for general use in buildings that are to be plastered, or in which partitions are likely to be shifted; and, as will be pointed out in Chapter VIII, the segmental arch is not as desirable as the flat arch when fire-resisting qualities are considered.

A special pattern of concave arch has also been employed, as illustrated in Fig. 12, in which the soffit formed a segment,

FIG. 12.—Terra-cotta Arch with Curved Soffit. Sibley Warehouses, Chicago.

while the top was flat. This type was used in the Sibley Warehouses, Chicago, where it was made extra heavy to carry warehouse loads.

Development of Column Fireproofing. — As long as iron was considered a fireproof material the columns were generally made of cast-iron, unprotected, with ornamental capitals and bases. The first attempt made to protect columns against the effects of fire was through the use of a double shell, one within the other, the intervening space being filled with plaster. This system was patented, and was required by the New York building laws in an effort to lessen the danger liable to occur from the use of unprotected columns carrying interior brick walls. The freezing up of the plaster filling, however, in some cases, caused the abandonment of this detail, and the air-space between the two casings was then relied on. Such provision may still be found in the New York building laws.

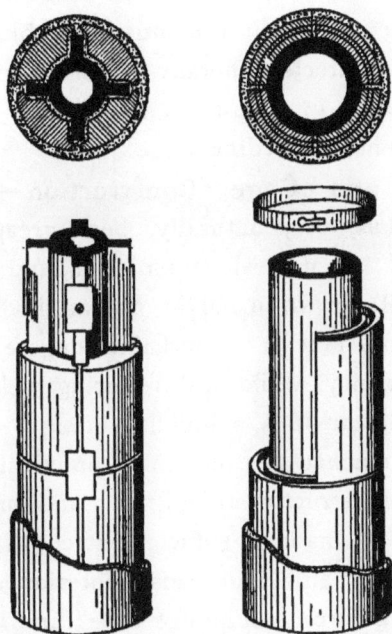

FIG. 13.—Early Forms of Column Fireproofing.

The first terra-cotta fireproofing for columns was used on columns of a cruciform section (see Fig. 13). Porous terra-cotta gores were used in sections about one foot long. They were secured to the metal columns by means of screws (which were tapped into the column flanges), and metal washers placed at the joint intersections as shown in the illustration. The whole surface was then covered with a plaster or cement finish. These gores were first employed in the Chicago Club House, where the shafts were plastered and the capitals were made of ornamental terra-cotta. Cast-iron flanged columns or Phœnix flanged columns were also fireproofed in the same manner in a large number of buildings during the years from 1880 to 1884.

In the U. S. government buildings built between the years 1880 and 1886 a system was employed consisting of 1½-in. terra-cotta slabs placed around circular cast-iron columns, the slabs being held in place by hoop-iron bands which were inserted in rabbets provided in the ends of the blocks.

After 1886, architects generally began to specify iron columns to be encased in terra-cotta blocks, and many of the present forms came into ordinary use.

"Skeleton" and "Cage" Construction.—The question of fireproofing has, very naturally, been greatly stimulated by the increased use of steel construction, and as steel buildings have steadily grown in height, the methods of fireproofing have developed to a remarkable degree in an effort to increase efficiency, while still reducing weight and cost. Previous to the year 1883, a building of eight or ten stories had very nearly reached the maximum practical height under the methods of construction then known and practised. The exterior walls were made of sufficient strength to carry their share of the floor- and roof-loads ; interior columns were either circular cast-iron or Phœnix type ; and fireproof floors were made of either brick or terra-cotta arches, the whole method being heavy and expensive as judged from present standpoint.

The design, in 1883, of a building in which iron columns

were built into the exterior masonry walls caused a most radical change in building methods, and as others followed, each improving on its predecessors, the system of replacing exterior load-bearing masonry walls with light curtain or veneer walls became established. The steel framework, originally introduced to carry vertical loads only, became gradually modified and strengthened as increased attention was given to the questions of lateral strength and stiffness against wind or other destructive forces.

The use of a metal framework is generally characterized as "skeleton construction," or "cage construction."

" 'Skeleton' is a term clearly descriptive of that type of construction to which it was first applied, a simple framework of columns and beams whose efficiency is dependent largely on the existence of exterior walls and partitions which brace the building and hold the framework in position, just as the utility of the human skeleton is dependent on the covering of sinews and muscles that hold the component parts together. On the other hand, the light framework of an ordinary wire cage bound into one compact unit is suggestive of an inherent strength and elastic resistance that renders any covering an incident rather than a necessity. 'Cage' is a term peculiarly descriptive of that type of construction represented by the most advanced and approved practice, a framework of columns and beams, spliced at the joints, riveted at the connections, stiffened by an efficient bracing of rods, portals, or gussets, that make it independently safe against any external force, leaving the thin and light exterior walls with no duty to perform except that of providing protection and ornamentation to the building." *

Origin of Skeleton Construction.—To Mr. W. LeB.

---

* "Wind Pressure in the St. Louis Tornado, with Special Reference to the Necessity of Wind Bracing for High Buildings," by Julius Baier. Trans. Am. Soc. C. E., vol. xxxvii.

Jenney, a prominent Chicago architect, is due the credit of having designed and executed the first example of skeleton construction. The departure from conventional methods was radical, and the successful completion of the building caused much comment and discussion, and soon resulted in the very general adoption of the system.

In 1883 Mr. Jenney prepared plans for a ten-story office building to be built in Chicago for the Home Insurance Company. In order to obtain a maximum of light for the offices, it was decided to reduce the width of the piers as much as possible. This suggested the use of iron columns in each pier to carry the floor-loads, thus relieving the masonry piers of these loads. The exterior walls were thus made self-supporting, except the spandrel portions, between the top of one window and the bottom of the window above. These parts of the walls were carried on girders placed in the exterior walls, extending from column to column.

Development of Skeleton Construction. — In 1885–86 the Rookery office building in Chicago was constructed. This building had self-supporting solid masonry walls and isolated footings. It was thought at this time that the supposed difference in the thermal expansion and contraction of iron or steel and masonry would constitute a serious objection to the employment of skeleton construction. Continuous columns of metal, 150 feet high, under a variation of 120° F., were thought to expand and contract so much more than the masonry envelope that a serious cracking of the brick- or stone-work was looked for. This suggested the carrying of the walls or piers themselves on the steel columns, story by story, thereby dividing the piers into single-story lengths. It was soon tried in succeeding buildings, the Tacoma Building in Chicago, fourteen stories in height, being the first complete type of skeleton construction.

As a matter of fact, the rates of thermal expansion and contraction of masonry and steel are so nearly equal that no disturbance need be feared from such causes, but these false fears were responsible for a great forward step in the question of high-building design—an improvement which otherwise might have been unknown for many years.

In the East, the Manhattan Life building in New York City was the first notable example of a building erected after the new methods. The project for this structure, with a moderate height of seventeen stories, caused much more comment at that time than has recently attended the erection of the Park Row Building of thirty stories.

Mill Construction.—Buildings constructed after the generally accepted methods of "slow-burning" or "mill-construction," have not been found sufficiently "slow-burning" in practice to warrant the extended use of this system. Mill-construction, as often practised, would be deservedly ridiculed by competent mill engineers, but even the best examples of this construction have proved a failure.

Mill-construction undoubtedly possesses some decided advantages. These consist in placing the timbers some distance apart, and using a thick flooring-plank, without ceiling of any kind. This woodwork will not ignite as readily as the thin joists and floors used in non-fireproof construction, and the absence of lath and plaster does away with the danger of concealed spaces between the floor and ceiling. The fire department is enabled to get at the fire more readily, because exposed, and the fire cannot be hidden, to break out somewhat later.

Objections to the use of mill-construction, however, lie in the general use of Southern or pitch-pine for the long and heavy timbers required in this system. This timber is cheaper and easier to obtain in the sizes called for, and it is also usually considered stronger than oak. But it is full of pitch, and, when once ignited, makes a hotter and fiercer fire than even the light joists and lathing of non-fireproof construction. It is, in fact, the most inflammable of building materials, and when fire is once started it spreads more quickly in all directions over the pitch-laden timber than was the case in the earlier construction.

The Boston fires in mill-construction buildings, the Jenkins Building in Pittsburg, and the Ely-Walker fire in St. Louis, have all shown that no advantage against loss can be expected from this type of building unless thoroughly equipped with automatic sprinklers.

But aside from the fire-resisting qualities, much trouble is experienced from shrinkage, deflection, warping, and dry-rot. These, added to a questionable security, tend to give mill-construction no ultimate advantage as to cost over a thoroughly fireproof construction depending upon a steel frame fireproofed with terra-cotta or concrete.

Concrete Construction.—Concrete has been used in buildings and in engineering structures for many centuries, but it is only in comparatively recent years that concrete has been employed in combination with metal tension-members. It is generally accepted in engineering practice that concrete, owing to its low tensile strength, must be used under compressive stresses only. This led to attempts at embedding iron rods, which were designed to furnish the necessary tensile strength which was lacking in the concrete. The first experiments in this line were made by Mr. Thaddeus Hyatt,* who made experimental beams of concrete with iron members introduced in a great variety of ways. These were tested by Mr. David Kirkaldy of London, and the

---

* See " Building Construction," by F. E. Kidder, p. 279.

results proved that such a combination of materials could be relied upon and placed in practical use.

The introduction of metal lath and wire netting into fireproof building construction soon suggested the use of these materials in combination with concrete, and such construction is now largely recognized as good engineering practice.

CHAPTER III.

FIRES IN FIREPROOF BUILDINGS.

DURING several years past a number of fires have occurred in so-called "fireproof" buildings, giving rise to much comment and criticism, both adverse and favorable, on the parts of the daily and scientific press. The buildings have generally been either those designed and constructed to resist fire according to the best or most popular system of fireproofing in use at the time of their erection, or those erected according to older ideas, or on a cheaper basis, and termed "fireproof" by the owner in an endeavor to increase their renting value.

As to this second class of structures, apartment houses, theatres, office and commercial buildings are constantly advertised as "thoroughly fireproof" when the construction in no way warrants such a designation. The owner has usually been content to erect a structure of iron or steel framework, brick walls, and possibly terra-cotta or concrete floor-construction, leaving external exposure and internal defects to care for themselves,—while the architect, perhaps often against his better judgment, has been allowed to give little attention to fire-resisting planning or detail, or to careful execution on the part of the contractor.

An illustration of the commotion which arises at times on the general subject of fireproofing, its efficiency or inefficiency, was afforded by a fire which occurred in November 1895 in a building of the above character in New York City. Starting in the Keep Building on one side of a street, the fire crossed the way and pretty fully destroyed the Manhattan Savings Bank building, "which, while far from representing the best fireproof construction, would ordinarily be called fireproof by *builders, landlords,* and the *public generally.*"* Note the criticisms which appeared shortly after, in which the entire system of fireproofing, as evolved by architects, engineers, and manufacturers, was assailed by the fire departments of our large cities.

A prominent official of the New York fire department was quoted as asserting: "I am prepared to declare from my experience that a building of brick and yellow-pine, in case of fire, is easier to manage, and the contents have more chance of being saved, than in a modern fireproof building." This reference to a modern fireproof building is further explained by the same official as follows: "The heat from the Keep Building on the opposite side of the street acted directly upon the *exposed ironwork* of the Manhattan building. The iron resisted the heat, that is, it did not blaze, but so far as the safety of the building was concerned it did something infinitely worse. It expanded under the heat and forced the ends of the beams and girders from their resting-places on the supporting piers. The result was inevitable ; the floors came down and brought with them the mass of fire-brick used as flooring, the floor-columns, and finally the roof."

Note the italicized words in the previous quotations, and the point, emphasized by the writer, of ignorance on the part of the owner and the general public, and misrepresentation on the part of the owner and interested builder will become more significant. Will any one who is at all acquainted with fireproofing call a building fireproof in which the ironwork is exposed, and open to direct attack from the flames ?

The wide range of improvements in the fireproofing field

---

* See *Engineering News*, vol. xxxv, No. 16.

of late years has caused the public to expect little short of infallibility in fireproof construction. The fact that fireproof structures can stand only in fireproof cities is usually lost sight of. The external fire hazard is no more to be neglected than the internal, and as long as highly inflammable buildings surround even the best of the fireproofer's efforts, the term fireproof is a misnomer. But how bitter nevertheless is the disappointment, and how unreasonable, often, the criticism of press and individual alike, when conflagration and the partial or even well-nigh total destruction of a really well designed and carefully built example prove the misgrounded faith in a structure claiming absolute fire-resisting qualities. Then becomes apparent what was entirely overlooked, that the word fireproof, as applied to the best of modern structures, does not mean immunity from all danger of fire, but an embodiment of the principles that have reduced the fire hazard, both interior and exterior, to a minimum, according to the best skill and judgment of the day.

Among the fires which have occurred in buildings termed fireproof, many have hardly been worth serious attention, because more was claimed for them than their construction warranted. Other fires, however, have taken place in buildings designed and built according to the best current methods.

A careful examination of past fires in this better class of structures would seem the best way of getting at data which are invaluable in the problem of fire-resisting construction.

Lumber Exchange.—In January, 1891, a fire occurred in the Lumber Exchange building in Minneapolis. This probably constituted the first really good test of fire-resisting methods. The building consisted of an old portion of nine stories, which had iron columns and girders, with wooden joists, this construction being called at that time "slow-burning," because all structural members, both iron and wood, were covered with terra-cotta. This old portion was enlarged by adding two more stories, making eleven in all, and also by annexing an entirely new portion of eleven stories. Both of these additions were made of the best fireproof construction then in use, consisting of a steel framework, with floor-beams spaced about 7 feet, centre to centre, and filled in with 5-in. hollow-tile floor-arches.

In the older portion of the building, the slow-burning construction burned for twenty-four hours. Of this construction, the iron columns that supported the iron floor-beams alone remained. These still partially retained their fireproof covering and, being little injured, supported the upper or newer stories of fireproof construction, which stood the great heat generated by the burning timbers of nine stories below without any serious damage.*

"Even those who have believed that buildings could be, and even were, made fireproof were surprised at the photograph taken from the sixth floor of the new building, looking through the opening into the old, showing the contrast as it has never been shown before in the history of fireproof construction. A factor that must not be forgotten in summing up the stability of fireproofing is the resistance to the weight and disintegrating effects of water. Tons of water must have fallen on these arches, rapidly changing their temperature from intense heat to cold, but neither weight, nor changes of temperature, nor the subsequent freezing (for the fire occurred in the coldest period of a Minnesota winter) seems to have in any way affected the arches. No greater contrast than is shown by this photograph can be well conceived of,—the destruction wrought where a 'good enough' policy was pursued, and the absolute safety where a common, practical system of fireproofing was employed." *

---

* See *The Inland Architect and News Record*, August, 1891.

This fire, although not in a fireproof building, is interesting as showing the comparison between slow-burning and tile construction in the same structure under test.

Schiller Theatre Fire.—This was an exposure fire due to the burning of a building of combustible construction near the Schiller Theatre in Chicago. The source of flame was some thirty feet distant from a court in the side wall of the theatre building, so that the fire was blown against the court wall, which consisted of an iron framework protected by six-inch vitrified tile blocks. As a result the outer faces of many of the tiles fell off, completely exposing portions of the iron frame. The fire attack was not a severe one, as adjoining brick walls were comparatively uninjured, but the fire entered the building by means of the windows and destroyed the contents and finish of several offices.

This instance clearly demonstrated the value of fire-shutters as a safeguard against exterior exposure, and showed the danger of using tile walls for exteriors. However, it has not deterred the erection of many very tall buildings since then with the walls exposed to severe exterior hazard. Another notable example was the burning of the Detroit Opera House in which case an adjoining steel-frame building of non-fireproof construction, but with exterior side walls of tile, was exposed to the flames. The tile cracked and fell off in large quantities.

Chicago Athletic Club Building Fire.—This fire, which formed the first severe test of modern fireproof construction, occurred on November 1, 1892. The building is nine stories in height, devoted exclusively to the purposes of the Athletic Club. The construction consisted of a steel frame with self-supporting exterior walls, porous terra-cotta "end-construction" arches, and partitions and column-covering of the same material. At the time of the fire the building had not been finished. The fire started in the gymnasium on the fourth

floor, a large room, two stories in height, with oak panelling
covering the entire walls and ceiling. Large quantitiès of
wood-finish for other portions of the building were also stored
in this room at the time of the fire, some piles being from
six to eight feet high. This large quantity of combustible
material naturally caused a very hot fire, which was after-
wards indicated by the resulting layer of charcoal débris in
this room, nine to twelve inches deep over the entire floor.
The flames escaped by the windows and doors and up the
stairway openings, entering all of the upper floors, consuming
all wood-finish, and completely destroying all plastering, electric wiring, piping, and the costly carved stone front above
the third floor.

The results of the fire included the twisting of floor-beams
in several cases and the bending of two steel columns, besides
considerable damage to the terra-cotta fireproofing material.
The tile partitions showed very poor resistance to streams
of water, over one-half of the column fireproofing dropped
off, and, in some cases, the under sides of the floor-blocks
fell off. The insurance adjusters' estimate of the damage
included :

Steel columns, beams, and girders $8,386.00
Tile arches.. 10,441.92
Tile partitions.................................... 3,173.82
Floor-strips 2,130.00
Concreting floors................................. 1,516.77
Grounds and rough frames for inside partitions. 4,260.00

The larger part of this damage, however, was due to the
details employed, rather than to the general method of protection, or materials used ; and as the faulty details constitute the
principal lessons to be drawn from this fire, they will here
be pointed out.

The ceiling of the gymnasium, of panelled oak throughout,
was fastened to 1-in. nailing-strips, which had been nailed to

the tile arches, thus leaving a space in which the air-currents and flame could freely circulate. This same construction applied to the panelled oak wainscoting, extending from floor to ceiling in the gymnasium, and to all corridors, which were finished in oak wainscoting to a height of five feet. Had this work been of incombustible material, or had a thorough system of back-plastering been used, a great amount of damage to the tile arches and partitions would have been prevented.

Another fault exposed in this test lay in the excessive amount of wood used in the floors, which consisted generally of two thicknesses—nailing-strips, spaced 16 ins. centres, which received a 6-in. by 1-in. fencing, on which was laid a 1-in. maple floor. These wood floors were entirely consumed, and in many cases the strips embedded in the concrete were completely burned through. This must have been a source of great heat.

But by far the greatest defect in fireproofing lay in the protection of the columns. This was accomplished by embedding nailing-strips in the tile covering to receive the oak finish around the columns. Between the floors and ceilings 2-in. by 4-in. wood strips were spaced about 3 ft. centres (the 4-in. face being left exposed), and the spaces between were filled in with terra-cotta blocks. As soon as the fire reached these wooden strips, they burned out and allowed the tile to fall. Such a combination of wood and tile cannot be too strongly condemned.

In repairing the building, most of the beams and even some that had been slightly bent were accepted as good enough, and the tile arches generally were accepted as strong enough for floor-support. Some damaged ceilings were repaired by means of expanded metal attached to the beams and damaged tile.

In a report made to the Building Committee of the Athletic

Association by Gen. Wm. Sooysmith and Mr. Isham Randolph, the following opinions were given :

" We have nowhere discovered that the metal portions of the building where the fireproofing held have been deformed or injured. And even where the fireproofing tile dropped off, from the burning out of the nailing-strips which supported them, the columns seem to have supported their loads without bending, except two on the eighth floor, owing, no doubt, to the fact that the greatest heat had been expended before the strips were so burnt away as to permit the tile covering to drop off. This building furnishes an assurance that was lacking before, that the metal portions of a building, if thoroughly protected by fireproofing properly put on, will safely withstand any ordinary conflagation if the quantity of combustible materials the building contains is not greatly in excess of that which enters into the construction of the building itself. In this instance we do not think that the fireproofing was properly bonded. The integrity of the building does not seem to be impaired, and it may be made as good as new by replacing the parts injured."

Pittsburg Fire.—Occurred on May 3, 1897, causing a total loss of about $2,500,000. This fire resulted in the partial destruction of three buildings (all, presumably of modern fireproof design), the loss of the entire contents of two of them, as well as the complete destruction of a non-fireproof building in which the fire originated. The relative positions of the buildings are shown in Fig. 14.

Buildings.—The Jenkins Building, occupied by the Jenkins Wholesale Grocery Co., with a stock of oils and other inflammable goods, was constructed entirely of wood, and is not of interest, therefore, as concerning fireproof methods.

The Horne Store Building, built in 1893, measured 120 X 180 feet. There were six stories and basement, without partitions (except for a mezzanine story in the rear), and with a

light well 22 × 50 feet extending from the first floor to roof
with an iron railing at each floor. Six elevators ran from the

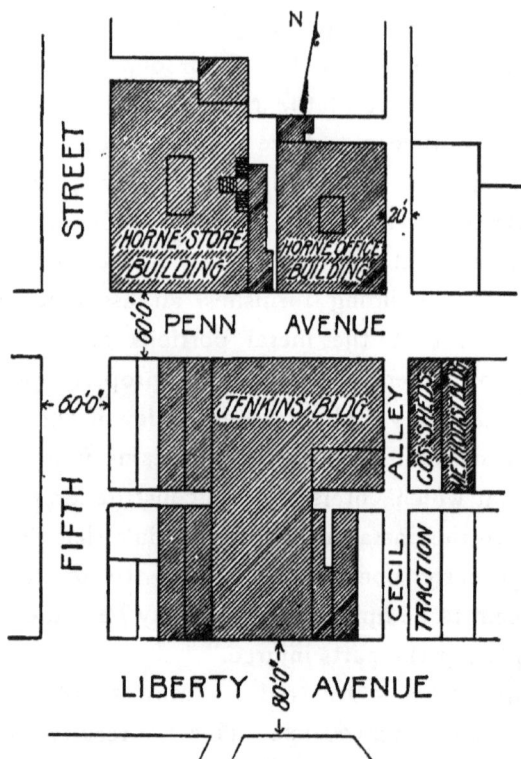

FIG. 14.—Plan showing Location of Buildings in Pittsburg Fire.

basement to the sixth floor—two being used for freight at
side and rear, and four for passenger service, grouped in connection with two stairways. All elevators were enclosed by
open grille-work, and the stairways were also open to the
various floors without enclosure of any kind. The floor and
column construction was as indicated in Fig. 15. The framing
for the floors was heavy, consisting of 24-in. box-girders,
with shelf-angles supporting 15-in. beams. The columns were
steel, standard Z pattern. The roof framing was made of
10-in. beams with no fireproof arches between, but covered

with light tees at right angles to the beams, on which 2-in. hollow book-tile were laid to receive the asphalt roof. A suspended ceiling was made of 1½-in. by 1½-in. tees, 12-ins. centres, carrying solid tile blocks 1½ inches thick. The floor-arches were 9-in. arches of hard-burned tile, side construction method, with webs about ⅝ inch thick. The column

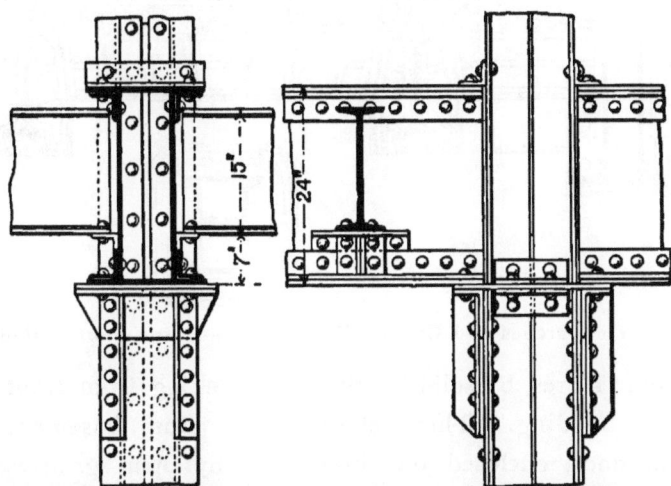

FIG. 15.—Details of Framing—Horne Store Building.

fireproofing consisted of 2-in. hard-burned tile ½ inch thick, with one air-space. See Fig. 16. The floor arches were covered with 4 inches of cinder concrete, in which were embedded the nailing strips, 14-ins. centres, to receive the hard-pine floors. All of the fireproofing was carefully applied. The front and rear walls were self-supporting, while the side walls were supported by the steel frame. The front windows were as large as could well be made, entirely unprotected by shutters. The rear wall had wooden shutters covered with tin. Over the stairways and passenger elevators was placed a compression tank, 6 × 25 ft., for operating the elevators; it rested on steel beams, and was enclosed with tile-walls and asphalt roof over. This entire building was occupied by

Horne's retail dry-goods store, fitted with shelving, counters, store fixtures and stock.

The Horne Office Building, built in 1894, was 94 × 136 feet, four stories and basement. The third and fourth floors were used for offices, being subdivided by 4-in. porous tile partitions. The first and second stories were subdivided

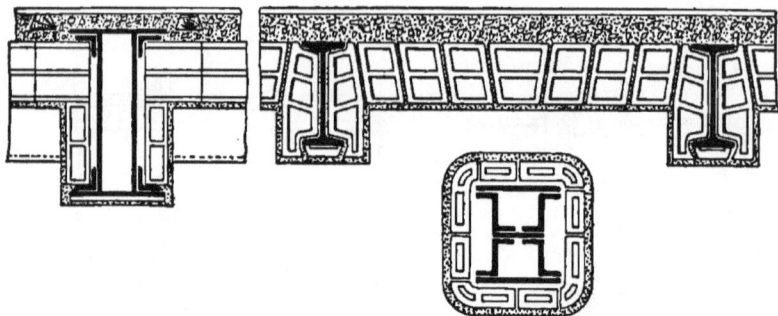

Fig. 16.—Floor-arches and Column Fireproofing—Horne Store Building.

into four stores, by solid partitions extending from front to rear of building. Three elevators ran from basement to second floor, enclosed on three sides by open grille-work. Another elevator and stairway extended from basement to fourth floor, both being in an enclosure surrounded on three sides with 4-in. porous tile. The floor-framing consisted of 15-in. beams, with double 20-in. beams as girders, supported by standard steel Z columns. The floor-arches were radically different from those used in the Store Building, being 9 inches in depth, but of end construction and of porous material, with ¾-in. webs instead of lighter webs of hard-burned material. See Fig. 17. They were covered with cinder concrete and wood flooring. The column fireproofing, the par titions, and the top-story ceiling were also made of porous tile, the column covering being 1-in. solid tile, instead of 2-in. hollow tile, as in the Store Building. The exterior walls were self-supporting. No shutters of any kind were used on this building.

The Methodist Building, built in 1894, was much smaller than the Horne Buildings. It was eight stories high, occupied by a book-store on the first floor, and by offices on the other floors. There were no interior columns, as 20-in. beams spanned the entire width of building, resting on columns in

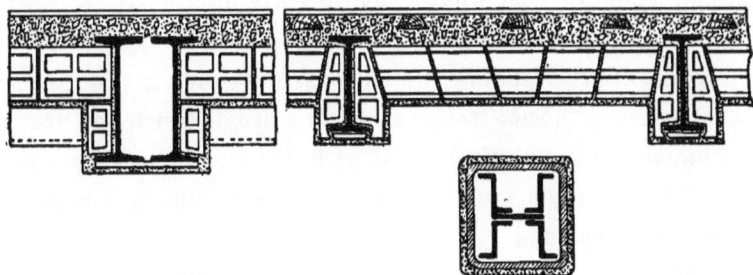

FIG. 17.—Floor-arches and Column Fireproofing—Horne Office Building.

the exterior walls. The floor arches were of the form of the Metropolitan System, but composed of a solid bed of Portland cement and furnace-slag concrete, about 8 ins. thick, with embedded wires carried over the tops of the 20-in. beams. The concrete covered the sides of the beams, and was flush on top with tops of beams. The partitions were made of 2-in. by 4-in. wood studding, covered both sides with wire lath and plaster, the spaces between the studs being left unfilled. The top-story ceiling was made of suspended wooden joists, covered with wire lath and plaster.

The Fire.—Originated in the Jenkins Building. It was soon seen that it would be impossible to save this structure, so the fire department directed all efforts to the saving of the Horne Buildings, but before much could be done the entire Pennsylvania Avenue front of the Jenkins Building fell, causing the flames to attack the front of the Horne Buildings with great suddenness and intensity.

The glass windows of the Store Building naturally offered but slight resistance, so that the entire contents were soon ablaze and by half an hour's time were entirely consumed.

The tank on the roof fell, carrying with it much of the steel construction, and causing great damage to the floors and columns, besides pushing out the adjoining wall for some inches. The vertical openings, light-court, elevators and stairways, undoubtedly produced a strong draught fed from the broken windows, which must have greatly intensified the heat and increased the damage.

In the Office Building the long narrow stores in the lower floors acted as horizontal flues for the fire, in much the same fashion as did the vertical openings in the Store Building, and all of the contents and non-fireproof constructional materials were soon consumed.

The flames also leaped across the alley from the Jenkins Building, and attacked the broad side of the Methodist Building, entering the offices upon the upper floors. This did not occur until the east wall of the Jenkins Building had fallen, as previous to that time, the iron shutters on the Jenkins Building side-wall had stayed the flames. As the stairs and hall of the Methodist Building were on the side away from the fire, the firemen had a good chance to fight the flames from the inside of each story. The contents of the building were pretty thoroughly consumed, but the constructional features were not very seriously affected, except such woodwork as entered into the construction.

Effects of Fire.—*Store Building.*—This suffered by far the greatest damage. The report of the engineers appointed by the adjusters * to examine the buildings and report upon the losses, estimated the total loss to the steelwork to be $18,530.33, or about twenty per cent. of the original value of the structural steel in the building. This amount included the material to be straightened and the necessary dismantling,

---

* Board of Engineers :   Gustave Kaufman, C.E., M. Am. Soc. C. E.; Emil Swensson, C.E., M. Am. Soc. C. E., Supt. Keystone Bridge Works; F. L. Garlinghouse, C.E., Chf. Eng. Jones & Laughlins, Ltd.

also the new material to be furnished and the erection expenses. Each and every beam, girder, and column in the building was subjected to close scrutiny, and a careful record made of the condition of all pieces. Beams or columns having deflections or bends of one-half inch or under were permitted to remain in place, but in order to determine to what extent the steel had been injured by the fire and what beams could be safely strengthened, typical pieces were selected and subjected to tests, and the opinion was reached that the fire had acted simply as an annealing furnace, and that the material had been improved rather than injured.

The brick fronts were seriously damaged by the cracking of the stone, and great damage was caused by the falling of the tank. The fireproofing throughout the building had to be replaced, excepting a salvage of 16⅝ per cent., a great portion of the loss being due to the falling tank. The tops of the tile arches were generally in good condition, save the cinder concrete filling, which was probably of poor materials when originally put in place. The ceiling-slabs of the floor-arches, of hard tile, were largely broken away, leaving hollow spaces in the arches, visible from the rooms below. The skew-backs and girder-casings were also badly broken.

The suspended ceiling over the sixth story was so bent and twisted that it was replaced entire. There was a lateral motion of the whole building amounting to two inches.

Office Building.—All of the woodwork was burned in places, especially where draught occurred, and in some cases nothing was left of the floor-sleepers. The steel construction was apparently little injured, so it was not uncovered for examination. The entire loss to the fireproofing, excepting the partitions, was 33⅓ per cent. of the entire cost. The loss on the partitions was 43 per cent. of the entire cost, being largely caused by the introduction of a wood strip to form a ground for the base-board.

The floor-arches were generally in good condition, the damage being confined mostly to the bottoms of the skew-backs surrounding the portions of the beams and girders that projected below the ceiling-line. The bottoms of the arches proper (porous) were not broken, as was so largely the case in the Store Building.

The column-covering generally remained intact. Some of the partitions remained standing, in spite of the burning out of the wood strip from column to column. Some partitions sagged when the door-frames burned out, and many fell down completely.

Methodist Building.—There was no room in which the woodwork was entirely consumed, showing that this structure did not receive the severe heat and consequent test of the other two. The firemen's efforts were also much more effective. The wood flooring was burned in only a few places, so that this can hardly be considered an adequate test of the fire-resisting qualities of the construction used. The showing, however, was most satisfactory, as the ironwork and the concrete floors were but little injured, and small expense was nceessary to make them entirely safe and serviceable.

The partitions were largely damaged, and offered but little resistance to the fire. The plastered ceilings fell off in large quantities, and the sixth floor ceiling suffered severely.

Structural Defects.—The following serious faults are to be noticed as having greatly increased the loss in the Pittsburg buildings, losses which were unnecessary, in that the faults could have been easily avoided.

The greatest mistake in the Store Building was in the roof construction, and in the supports for the tank used for elevator service. This tank was 6 ft. in diameter and 25 ft. long, weighing, when filled, about 52,000 lbs. It was supported by four 15-in. beams, which were in turn supported by the roof-girders. The entire roof, as before described, was protected by the suspended ceiling only, as were the beams under the tank, and, "judging from the warped and weak condition of the remaining ceiling-tees in the western part of the building, a fair surmise would be that the ceiling underneath the tank, which was located right beside the elevator-shaft, gave way under the heat, which was, no doubt, fiercer there than at any other place in the building. The fall of the ceiling exposed to the action of the heat the unprotected columns of the sixth story, between the ceiling and the roof, and also the roof-beams and those supporting the tank. These no doubt yielded on account of their loads and the action of the heat, and thus permitted the heavy tank to fall. The tank in falling carried with it to the first floor such columns and beams with which it came in contact, and these in turn dismantled a large number of beams and girders to which they were connected." *

Also, "the damage to the steel in the Store Building, outside of damage by the falling of the tank which tore away the fireproofing, would not have exceeded 5 per cent. of the entire cost of the same." †

Another serious structural defect in this building was in the use of wooden lookouts to support the copper cornice. This cornice was a total loss, but had steel brackets been used, fireproofed, as is usual in buildings of careful design, there would have been little or no loss.

In the Office Building, the principal structural mistake was in the use of wooden nailing strips for the attachment of the base-board. This must have been entirely unnecessary, as porous tile will hold nails nearly as well as wood. Aside from this, the partitions were nearly as good after the fire as before, but owing to this defect, nearly all had to be taken down.

In the Methodist Building as well, the principal defect lay

---

* From Report of Board of Engineers.
† Appraiser's Report.

in the partitions, made, as before stated, of 2-in. by 4-in. studs, wire-lathed and plastered. " The fire seems to indicate that plastering will not prevent the heat of a great fire from char-ring a wooden frame behind it, even if it does not come off and expose the wood, and such work is not fit to be reckoned as fireproof construction." *

Livingston Building.—This fire occurred on January 18, 1898, in New York City. The fire is of interest principally as indicating the action of metal-covered doors under fire-test, and the ability of such doors to confine a severe fire to the quarters in which it originated, without serious damage to the other floors of the building.

The building is a warehouse eight stories high, with steel framework, and floors of 12-in. end-construction hollow tile, projecting below the bottom flanges of the floor-beams, but not enclosing them. The beam-flanges were covered with expanded metal and plaster. The steel columns were pro-tected by a close wrapping of expanded metal, upon which was placed a coating of 3 to 1 Portland cement-mortar. The stairway and elevator-shaft were separated from the rooms on the various floors by a 2-in. partition built of metal studs, expanded metal lath and plaster. The doors and door-casings in these partitions were made of wood, covered with sheet steel over an intermediate lining of sheet asbestos. The two rear sides of the building had all windows protected by sheet-iron shutters.

The fire broke out on the fourth floor, occupied by furriers, and the combustible contents produced a heat so intense that the brass connections to the radiator valves were melted off. The windows were broken by the flames, and, where unpro-tected by shutters, the fire reached up externally and entered the windows of the floor above. The loss included the de-struction of the contents of the fourth story by fire, and dam-

---

* Corydon T. Purdy in Transactions Am. Soc. C. E., Vol. XXXIX.

age by water on the fifth and third stories, the latter being very slight, the result of the leakage through the ceiling. Had the windows on the street front been protected by sheet-iron shutters, as were those in the rear walls, it is probable that the upper floors would have suffered little or no damage.

The structural damage was comparatively slight. One of the steel girders suffered a slight deflection, the plastered ceiling fell off in places, and the column fireproofing gave way where the fire was most severe. The original cost of the structure was about $100,000, and the architect's estimate of repairs and reconstruction was $5,800.

This test quite effectively demonstrated the value of metal-covered doors in fireproof partitions. In the case in question, the doors were closed and intact upon the arrival of the fire-men, thus shutting off the flames from the hallways.

Another important point was the manner in which the column fireproofing of expanded metal and plaster acted under fire and water. This wrapping finally gave way, but it was sufficient largely to protect the steel columns from injury, in spite of a heat of two hours' duration.

The value and importance of fireproof shutters were here demonstrated, as in earlier and later fires.

Vanderbilt Building.—On February 11, 1898, the 15-story skeleton construction Vanderbilt Building, in New York City, was subjected to a severe exposure fire, through the burning of the Nassau Chambers, an adjoining 7-story non-fireproof building. One wing of the latter building was only 40 ft. away from the wall of the Vanderbilt Building, which had nine windows on each floor facing the fire. The flames from the burning building naturally broke in the windows, which were all provided with iron shutters, *none of which were closed*. The adjacent offices were completely gutted, but the damage was confined to the woodwork, contents, plastering, and windows. Great difficulty was experienced by the firemen in carrying their hose up the narrow crooked stairway, and several of them were overcome by the heat and smoke. This suggests the desirability of constructing fireproof stairways which shall be easily accessible, as free from turns as possible, and arranged so as to prevent the presence of smoke in large quantities. Adequate elevator service for the use of firemen is still more desirable.

The advisability of providing fire-resisting shutters for windows has already been pointed out, but the fact that such shutters were in place, but unclosed, is a matter of carelessness only, and open to more criticism than their entire absence.

Home Insurance Building Exposure Fire.—This fire, like the Pittsburg disaster, has been of special interest and value to all who are interested in fire-resisting methods. The building was undoubtedly well designed against internal hazard, and had the fire originated from within, there can be little doubt but that it could and would have been localized to the floor on which it occurred, and extinguished by means of the fire appliances provided.

Less thought, however, had been given to external hazard, and as the origin and extent of the disaster were due to external causes, the loss is largely attributable to this neglect.

On the night of December 4, 1898, a severe fire broke out in the 5-story building on the southwest corner of Broadway and Warren St., New York City. This structure was immediately adjacent to the north side of the Home Building, and was of ordinary brick and wood construction, occupied by a clothing establishment, and by offices on the upper floors.

The Home Building, erected in 1893, is of modern steel-frame construction, fifteen full stories in height, with a sixteenth story on the roof at the base of a pyramidal tower which reaches

to a height of 260 feet above the curb.   Fig. 18 gives the
typical plan of the upper floors.   It will be noticed that there
are two courts or light shafts, the one to the south being

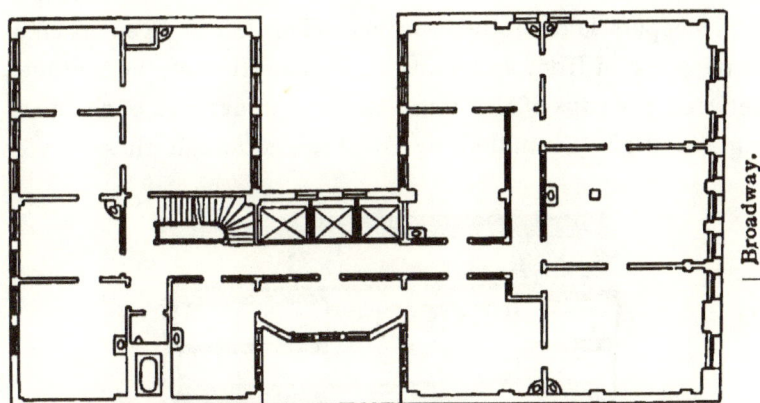

FIG. 18.—Typical Floor Plan—Home Life Insurance Building.

10 × 20 feet in size, looking out upon the 14-story steel-frame
building of the Postal Telegraph Company, and the court
in the north wall, about 20 × 24 feet in size, running from
basement to roof, on which the non-fireproof building abutted.

Both shafts are faced with white enamelled brick.   The two
side walls of the north court have four windows each on each
floor, with narrow mullions between, while the remaining or
inner side of the shaft has two smaller windows at each story,
opening into the elevator-wells.   Besides these court windows
the main north wall had two windows on each floor, overlook-
ing the corner building.

All exterior walls are carried on the steel-frame, except the
front wall, which is of white marble, self-supporting.   The
steel-frame consists of H-shaped columns, made up of plates
and angles, and plate-girders running transversely across the
building, knee-braced to the columns to resist wind pressure.
Nine-inch floor-beams are used in the front portion of the
building, and 12-inch beams in the rear portion, spaced about

4 ft. 6 ins. centres, and attached to the girders by angle-brackets. The floor-arches are 10-in. hard tile, side construction, with skew-back protection flanges. The wooden floors consisted of two thicknesses of $\frac{7}{8}$-in. boards, nailed to 3-in. $\times$ 4-in. sleepers laid on the I-beams. This left an open space of some 4 ins. in front and 7 ins. in rear portion of the building, between the tops of the arches and the under side of the flooring, see Fig. 19, unfilled by concrete, although these spaces

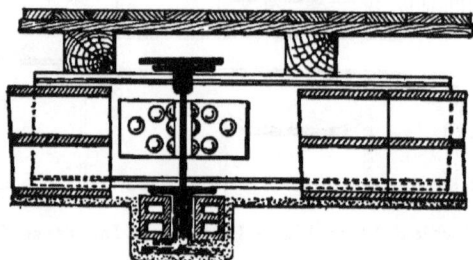

FIG. 19.—Details of Floor Construction—Home Life Insurance Building.

were supposed to be divided at intervals by concrete stops, built up solid from floor-arch to wooden flooring. This construction, however, left the top flanges of the beams and girders exposed: about one inch in the case of the beams and two inches for the plate-girders in the front portion of the building, and more in the rear portions where the beams were deeper, but where the floor-arches were of the same thickness as in front.

The lower flanges of the girders, where they projected through the ceilings, were protected by tile blocks above the flanges, while the knee-braces, connecting the girders to the columns, and the lower flanges of the girders were wrapped with expanded metal lath, covered with a thick coat of plaster. The walls were lined with 2-in. ribbed porous tile furring-blocks, and the columns were encased with the same material, thus giving two air-spaces. The partitions were of 4-in. porous tile, the upper 4 ft. in the hall partitions being filled by windows.

The fire occurred during a very severe northeast gale, which blew the flames, escaping from the windows and roof of the clothing establishment, directly against the north walls of the Home Insurance Building. "Although subjected to an intense heat, the building did not take fire for some time. The firemen entered the Home Building and endeavored to protect it with hose-streams from their engines and from the fire-pipes in the building," but about an hour after the fire broke out, the roof of the corner building fell in, and a great mass of flame was thus concentrated upon the open court, which acted as a great flue for the upward rush of fire. The windows of the upper floors gave way, thus admitting the fire almost simultaneously to the different stories.

The greatest damage was done from the eleventh floor up, being greatest in the rooms adjacent to the court, and decreasing as the fire spread inward and around to the centre of the opposite side, or to the court adjoining the Postal Telegraph Company's building. From the eleventh floor down the damage gradually decreased, until at the seventh floor it was due principally to smoke and water. Up to the eighth story the firemen were able to work successfully, while above that level the pressure and volume of water obtainable with their fire-apparatus were not sufficient, and the intense heat drove them from the corridors and prevented the use of the streams provided by the stand-pipe system in the building.

The fire was also communicated to the Postal Building by means of the south light-shaft, but the damage was mostly confined to the adjacent walls and windows. Some of the rooms in the upper stories were burned, but without serious loss.

Structural Damage to Home Building.—The main structural features of the building were pretty effectively preserved, and the value of the fire-resisting construction was amply demonstrated. The floor-arches stood their test remarkably well, and served as almost perfect barriers to the passage of fire from floor to floor. With few exceptions, the terra-cotta arches required little or no repairs. The principal exceptions were on the 10th and 15th floors. In the former case the failure of the arch which broke through was due to the falling of a safe, probably caused by the burning through of the wooden flooring between the I-beams, thus allowing the safe to fall several inches through the free air-space onto the arch, and shattering the terra-cotta blocks. Had this space between the tile arches and the wooden flooring been filled with cinder concrete, it would have largely prevented the burning of the wooden flooring and sleepers, which were almost entirely consumed above the seventh floor.

This air-space was also responsible for the failure of the terra-cotta arch on the fifteenth floor on the west side of the light-shaft. In this case an entire panel of floor-arch fell, and the remaining arches between it and the stairs sagged so badly as to require removal. This was caused by the deflection of the 12-in. I-beams, caused by the exposure of the upper flanges, after the flooring had been burned away. That more similar cases did not occur can only be attributed to the limited height of the exposed metal. A few other floor-arches sagged, and the lower webs of individual blocks cracked off, but the condition on the whole was very satisfactory.

The lower flanges of the floor-beams, protected by the skew-back flanges, were, as a rule, almost perfect ; while the girders which projected below the ceiling were in many cases bare, the wrapping of expanded metal and plaster either having dropped off completely, or hanging loosely to the girder. This fire plainly shows that plastering on wire lath or netting will drop off, except such portions of the plastering as is pressed through on the reverse side of the lath or netting. Appearances indicated, however, that the protection had stood long enough to preserve the metal from serious injury.

The column-covering of porous tile saved the columns from

injury in every case, although in places it was knocked off in patches. In many instances the cement bond of the joints was broken, and resetting was necessary.

Just how much damage was done to the partitions by fire and water, or what injury was done by the firemen who knocked many of them down to get at the flames, it would be hard to say. The burning of the wooden doors and windows in the partitions, and their casings, were probably responsible for much damage; and the common plan of locating such partitions to suit tenants, placing them indiscriminately over the wooden floors after the completion of the building, with insecure attachment to floor and ceiling, adds instability (upon burning away of floor-boards) to what must be admitted as being the weakest feature of fire-resisting methods—namely, block partitions in general. The methods of partition construction must certainly be greatly improved to make them fire- and water-resisting. In the twelfth story some partitions made of plaster on a framework of small angle-studs, covered with expanded metal, the total thickness being 2 ins., remained in position, though they were badly distorted. Their insufficiency was amply demonstrated, but had the force of fire-hose been added to the heat, the result would undoubtedly have been still worse.

The structural damage also included the cracking and spalling of the marble front to such an extent that it had to be rebuilt above the eighth floor. Portions of the cornice and balcony and other ornamental marble-work fell to the street, and other portions were rendered unsafe and had to be removed or shored up. The rear, side, and court walls stood the test well, emphasizing the superiority of brick and terra-cotta over stone as a fire-resisting material. On the tower, the red roofing tiles were not injured a particle. The iron framework of the elevator shaft and the stairway endured the fire remarkably well.

Various structural defects which existed in this building have been pointed out, and these assume equal importance against either internal or external fire hazard. But criticism may also be made regarding the absence of any fire-resisting shutters to the windows in the courts and the side wall. Commenting upon this point, the *Engineering News* of December 8, 1898, says: "From what has been said it will be generally admitted by those competent to judge, we think, that the value of its fireproof construction was proven by the Home Life Insurance Co.'s Building. Its weak feature was the unprotected window openings in the light-shaft. To this single instance of neglect, we believe, the damage of the building by fire was entirely due. Indeed, we have no hesitation in saying that had the light-shaft windows been protected by good fireproof shutters, and been set in metal frames and sash, the flames would never have entered the building with the firemen behind with hose to keep the shutters cool. The building was well fireproofed in every respect except to withstand an exposure fire, and with the tinder-like clothing store by its side an exposure fire was the one which the builders had the most reason to expect. As it was, the building kept the fire confined so that its spread was easily prevented under conditions of high wind and bad facilities for fighting flames so far above the ground."

CHAPTER IV.

TESTS OF FIREPROOF FLOORS.

General Requirements.—Considered from the owner's, architect's, builder's and underwriter's standpoints, the following requirements may be stated as the principal ones for an acceptable fireproof floor:

Strength and rigidity under static and moving loads;

Ability to resist shock due to falling bodies;

Permanency under ordinary conditions and service;

Minimum of dead weight in the construction itself;

Simplicity of construction, not involving skilled labor;

Minimum of cost, using cheap, abundant materials.

In addition to these requirements, the floor must also possess satisfactory fire- and water-resisting qualities, and be able to stand successfully severe heat for a considerable time, followed by applications of cold water, and this under the action of heavy concentrated loads, without material injury or excessive deflection.

To determine the ability of floor systems to meet these requirements, careful and scientific tests have been found to be the only practicable method. Various individual and collective tests have been made on the fifty or more different systems of floor construction now on the market, the more important ones being as follows:

Denver (Colo.) Tests.—Made in Denver, Colo., in December, 1890,* for the Denver Equitable Building Co., under the supervision of Messrs. Andrews, Jacques, and

---

* See *The American Architect and Building News*, March 28, 1891.

Rantoul, architects.   This series of public tests formed one of the earliest as well as one of the most valuable series of competitive tests undertaken.

When bids for executing the fireproofing contract for this building were opened, it was found that three competitors had figured on the work, two of whom estimated on furnishing floor-arches of hard tile, side-construction, and the third, at the highest price, figured on furnishing the arches of porous tile on a new principle, now well known as the "end-construction" method.   In order that the relative qualities of the different systems might be compared, the architects decided to institute a series of tests, the conditions including

A—a still-load, increased to failure of arch;

B—shocks, repeated to failure of arch;

C—fire- and water-test, alternating until arch was destroyed;   .

D—continuous fire of high heat, until arch was destroyed.

Twelve arches in all were tested—three, or one for every competitor, under every condition.   The arches were 10 inches in depth, built between I-beams 5 ft. centres.

Still-load Test.—The "Lee" or end-construction arch, sustained a final load of 15,145 lbs. for two hours, the deflection being .065 of a foot.   The heaviest load sustained by a side-construction arch was 8,574 lbs.

Drop Test.—The blows were given by dropping upon the arches a piece of Oregon pine, 12 inches square and 4 feet long, weighing 134 lbs., from a height of 6 feet.   Both of the side-construction arches were shattered at the first blow, while the end-construction arch stood up to the eleventh drop from a height of 8 feet.

Fire-and-water Test.—One of the side-construction arches was destroyed by three applications of water with a fire-temperature of 1,300 degrees, the other being very badly

shattered after fourteen applications of water. The end-construction arch was given eleven applications, and at the end
of twenty-three hours was practically uninjured, as it required
eleven blows from the ram used in the drop test to break the
arch down.

Continuous-fire Test.—Of the two side arches, one failed
completely after a continuous fire of twenty-four hours, while
the other arch stood, but was unable to carry a load of 300
lbs. per square foot. The end-construction arch supported a
weight of bricks of 12,500 lbs. on a space 3 ft. wide in the
middle of the arch, after a continuous fire for twenty-four
hours.

The Denver tests are described here at some length, because they are of particular interest and value as showing that
manufacturers and architects continue to produce and use
materials and forms which have been proved vastly inferior;
and the results are not appreciated until these materials and
forms have been tried in severe fires in finished and occupied
buildings, and been found wanting.

The results of the Denver tests have never been ques-
tioned, as they have been rendered all the more emphatic by
testing two separate sets of hard tile side-construction arches,
with nearly identically poor results, as compared with the
third satisfactory test of porous-tile arches of end-construction. Two important facts were here established beyond
question—namely, that hard tile is brittle, unable to stand
fire- or water-tests, and is therefore very inferior to porous
tile; and, secondly, that the side-construction method of tile
floor-construction cannot be favorably compared with the
end-construction type.

St. Louis Tests —In 1891–92 Messrs. J. B. Johnson and
Edward Flad, Members Am. Soc. C. E., made a series of
tests for floor-arches in the Wainwright Office Building in
St. Louis, Mo. These consisted of concrete arches sprung

between I-beams 8 ft. centres; the arches being 15 ins. deep at the haunches, and from 2 to 4 ins. deep at the centre. The concrete was made of one part by volume of cement to four or five parts crushed coke or cinders. A flat ceiling was also constructed, entirely separate from the floor-arch, made of tile $\frac{3}{4}$ in. thick, hung from 2 × 1$\frac{1}{8}$-in. iron bars spaced 7 to 14 ins. centres, and resting on the bottom flanges of the I-beams.

Five kinds of tests were used:

A—Uniform-load test, to destruction.

B—Concentrated-load test, to destruction.

C—Dropping-test, to destruction.

D—Fire-test, to destruction.

E—Fire- and water-test, to destruction.

Hydraulic pressure was used in the load-tests, secured by water-pressure introduced between two diaphragms spread over the arch. The fire-test was made by using eighty-four gas-burners, placed immediately below the ceiling construction. The temperature was indicated by a thermo-electric pyrometer.

Arch No. 1, 2$\frac{1}{8}$ ins. thick at the centre, broke at the centre under a load of 16,400 lbs., equivalent to 559 lbs. per sq. ft. of loaded area, or 417 lbs. per sq. ft. of whole area tested.

Arch No. 2, 4 ins. thick at centre, broke at the centre under a load of 39,360 lbs., this being equivalent to 1,342 lbs. per sq. ft. of whole area. After breaking, the load was still carried, the arch having sunk 1$\frac{1}{4}$ ins. in the middle.

Arch No. 3. The load was placed on a side portion only of the arch. Fracture occurred under a load of 12,027 lbs., or 768 lbs. per sq. ft. of loaded surface, or 650 lbs. per sq. ft. for one-half of the whole area.

Arch No. 4, 2$\frac{3}{4}$ ins. thick at the crown, was subjected to a drop-test. A pine stick weighing 134 lbs. was dropped twice from a height of 6 ft., then up to the seventh blow was dropped 8 ft., and then reduced to 6 ft. again. Thirteen blows were required to destroy the arch completely, the previous blows, from the seventh to the tenth, having caused cracks and made pieces of the arch fall out.

Arch No. 5 was subjected to a fire-test applied to the ceiling, as this was supposed to furnish the fire-resisting medium. The ceiling successfully withstood a temperature of about 1,500 degrees for six and one-half hours, being depressed about one inch at the centre; but, after cooling, it resumed its original position.

Arch No. 6 was the same arch as No. 5, tested for fire and water effects. The stream from a $\frac{3}{4}$-inch hose was turned on to the bottom surface of the ceiling while still at a temperature of about 1,500 degrees. A few cracks appeared, but no serious damage was done. The arch was then reheated and drenched again, and subjected to the same operation for a third time, after which the construction was found to be practically uninjured.

Trenton (N. J.) Tests.—In 1894 tests were made at Trenton, N. J., to determine the relative values of the flat-tile arch, the segmental-tile arch, and the Metropolitan type of suspended flooring.

Test-arch No. 1 consisted of a flat-tile arch, side-construction, 10-in. blocks, span 4 ft. 11$\frac{3}{4}$ ins. The load was applied at the quarter and three-quarter points of span. Under a total load of 6,689 lbs., or 301 lbs. per sq. ft., the joints commenced to open at the under side of the keyblock. Under a total load of 8,599 lbs., or 387 lbs. per sq. ft., the skew-backs were crushed and the arch was entirely destroyed.

Test-arch No. 2 was a segmental side-construction tile arch, span 5 ft. 2 ins., rise 6 ins., thickness 6 ins. Under a loading of 6,942 lbs., or 301 lbs. per sq. ft., the joints commenced to open, and at 9,492 lbs., or 411 lbs. per sq. ft., the skew-backs were broken as before.

Five other tile arches were tested, one of which, No 4, sustained a load of 20,140 lbs., or 839 lbs. per sq. ft. The arch was made of 10-in. blocks, 5 ft. 6 in. span. Failure occurred by the breaking of the horizontal webs of the blocks next to the skew-backs.

For the test of the Metropolitan floor, the slab used consisted of a mixture of five parts plaster of Paris by weight to one part wood shavings, with sufficient water to bring the mass to the consistency of thin paste. The wires used were spaced about one inch centres, hooked over the beam-flanges. The arch-slab was 4 ins. thick, span 5 ft. A section of the arch 2 ft. $6\frac{1}{2}$ ins. long was cut out on the completion of the sample, and the whole of this detached area was loaded for the test. Cracks appeared under a load of 735 lbs. per sq. ft. Under a load of 1,200 lbs. per sq. ft. the deflection was $1\frac{17}{32}$ ins.

A test for impact was also made on a section of the Metropolitan floor, also a fire-test, for which an arch of 5 ft. 6 in. span was loaded to 200 lbs. per sq. ft. uniform load, with fire sustained directly beneath the arch for four hours. The results were considered highly satisfactory.

In addition to the foregoing, most of which combined tests for fire-resisting qualities with ability to carry loads, the following tests of floor-arches have been made, among others, for carrying-strength only.

Tests by Fr. Von Emperger. — Made in January, 1895, to compare the relative strength of a flat side-construction arch with a segmental arch of tile.* See Figs. 20 and 21.

Test-arch No. 1 failed without appreciable deflection

---

* See " Hollow Tile Floors, Past and Present," by Fr. Von Emperger. Trans. Am. Soc. C.E.: vol. xxxiv, No. 6.

under a load of 10,600 lbs. Test-arch No. 2 failed under a
load of 19,000 lbs. Reducing these figures to a uniformly

FIG. 20.—Terra-cotta Arch used in Tests.

FIG. 21.—Segmental Terra-cotta Arch used in Tests.

distributed load, as specified by the building laws, the results
gave :

Test No. 1—Total breaking-load, 856 lbs. per sq. ft.
Test No. 2—Total breaking-load, 1,616 lbs. per sq. ft.

From these tests Mr. Von Emperger suggested the employment of tile arches with curved ribs and strong arch sec-

FIG. 22.—Proposed Form of Terra-cotta Arch.

FIG. 23.—Proposed Form of Terra-cotta Arch.

tion, as in Figs. 22 and 23. The curved ribs only would be
relied upon to take the strains, and no reliance would be

placed upon the joints. The least dimensions of these ribs could be distinctly specified according to the load, while the remaining portions, to form a flat ceiling, could be reduced in thickness and consequent weight.

Tests by Geo. Hill.—Made in 1894–95 on a series of Melan and terra-cotta arches.* Previous to these tests, fireproof floors had generally been treated to one of three forms of loading: 1, pig-iron; 2, piles of brick; 3, hydraulic pressure of very limited capacity. Where piles of pig-iron or brick were used, the great mass of load could not be readily handled, and in all cases difficulty was experienced in properly observing the way in which the arches acted immediately before failure. Also, the practice of the manufacturers of using well-bonded piles of brick or other material served largely to relieve the centre of the floor-arch, due to the arch-like action of the load itself.

Mr. Hill, therefore, had constructed a portable, self-registering hydraulic machine, capable of testing, to destruction, building-floors of any ordinary span, and made of a pattern which would make it possible to examine carefully the arch under test, up to the point of failure.

Melan concrete arches were made by placing three ribs of either 3-in. tees or 4-in. I-beams between 12-in. I-beam skew-backs. These ribs were curved to an arc of about 8 ft. radius, and rested on the lower flanges of the 12-in. beams, with thin steel wedges driven between the webs of the I-beams and the curved ribs. A wooden center was then placed flush with the under side of the curved ribs, and concrete filling poured in, consisting of a coarse 1–2–4 mixture for two-thirds of the bulk, and a 1–2 mixture for the balance or upper third of the arch-ring. Ten tests were made, the most important being as follows:

---

* See " Tests of Fireproof Flooring Material," by Geo. Hill. Trans. Am. Soc. C. E., vol. xxxiv, No. 6.

Span. Feet.	Rise. Ins.	Width. Feet.	Areas. Square feet.		Loads. Pounds.			Deflection.	Thickness.	Material.	Remarks.
			Total.	Loaded	Total.	Per sq. foot of area.	Per sq. foot loaded,				
6.0	8.53	4.0	24	1.00	40,000	1,667	40,000	0.43	3	Tees.	No. 81
6.0	7.28	3.0	18	1.25	55,000	3,055	44,000	0.47	4	I's	No. 82 } Same
6.0	7.28	3.0	18	1.25	58,750	3,260	47,000	0.59	4	I's	No. 83 } arch.
7.0	7.52	6.0	36	1.00	55,000	1,528	55,000	0.58	4	I's	No. 84
6.0	7.28	6.0	36	1.00	70,000	1,945	70,000	0.10	4	I's	No. 87

Arch No. 87 was loaded eccentrically, 1 ft. 7½ ins. from one skew-back, over the central arch beam. At 70,000 lbs. the clamps on the testing-machine broke, without failure of arch.

Tile-arch tests included arches of both side- and end-construction methods, and of both porous and hard terra-cotta. The following table gives the results of tests made on arches built for the purpose, about three months old, laid up in strong Portland cement.

Test No.	Span. Inches.	Width. Inches.	Rise. Inches.	Depth of skew. Inches.	Total arch area. Square feet.	Loaded arch area. Square feet.	Total load. Pounds.	Type of skew.	Equivalent distributed breaking loads.			Remarks.
									Porous terra-cotta, end-construction.	Hard terra-cotta, end-construction.	Hard terra-cotta, side-construction.	
91	60	48	3.5	6.0	20.0	3.00	13,750	Protection	.....	688	.....	Crushed on arch lines.
92	65	50	6.0	10.0	22.6	20.00	57,500	Plain	.....	2,545	.....	Unbroken, mortar frozen.
93	60	36	6.0	8.0	15.0	0.82	10,000	"	.....	.....	1,336	Sheared by edge of load.
94	60	36	5.0	8.0	15.0	4.98	5,700	Protection	.....	.....	381	Sheared by skew.
95	60	35.2	5.0	7.5	14.7	0.86	11,250	"	.....	1,534	.....	Crushed porous terra-cotta skew.
96	60	36.5	5.0	7.5	15.2	0.88	13,000	"	1,714	.....	.....	Crushed porous terra-cotta skew.
97	60	36	7.5	10.0	15.0	4.98	13,500	"	.....	....	900	Crushed on top by key.
98	60	37	8	10.0	15.4	4.98	14,500	Plain	.....	.....	943	Side skew failed.
99	60	41	10	12.0	17.1	0.88	16,400	"	.....	1,920	.....	Porous terra-cotta skew.
100	60	38.25	7	8.0	15.8	0.88	14,500	"	1,838	.....	.....	Porous terra-cotta skew.
101	60	38.25	7	8.0	15.8	0.88	15,750	"	.....	1,994	.....	Porous terra-cotta skew.

In the above table the load per square foot of arch for concentrated loads has been doubled for the purpose of comparison between distributed and concentrated loads. The results show the strength to be about the same for the end-construction arches of either porous or hard material, and prove that this construction is superior to the side-method arches.

In addition to the tests given above, a considerable number were made on arches laid without mortar joints, but the results showed great variations, and they are of doubtful value because in ordinary practice mortar of some kind is always used. Mr. Hill draws the following conclusions from his experiments:

" First, That the side construction arch requires a skew in which the inclined member on the line of the arch-thrust runs back to a solid support against either the fillet or the flange of the skew-back beam.

" Second, The end-construction arch requires good mortar and reasonably good bedding to make the joints which transmit the pressure adequate to their duty, the weak point being the mortar in the end joints.

" Third, When well bedded the end-construction arch is very much the stronger."

Doubtful Tests.—Many other tests have been made at various times and places, most of them being for load resistance only, and very few to determine fireproofing qualities; but nearly all such tests have been made by different individuals, often interested in the outcome, seeking to obtain the most favorable conditions and results for some particular product. Such tests have differed widely in time, place, conditions, and ends sought after, thus absolutely precluding any possibility of fair comparison.

Tests Made by the New York Building Department.— The lack of exact, impartial knowledge respecting many of the various systems of fireproofing in use led the Superintendent of Buildings in New York City, Mr. Stevenson

Constable, to undertake a series of exhaustive tests to determine the comparative merits of the more important methods of floor construction, and in 1896 he asked a number of companies to submit test samples of their respective constructions, such tests to be official and uniform in requirements. A direct comparison could thus be made, at once thorough and impartial.

Vacant lots for the test structures were secured in New York City, and the tests were conducted by the officials of the New York Building and Fire Departments. The kilns or test-houses were built according to plans prepared by the Building Department, and the various floor companies then built their floors over these chambers (each floor being about 14 ft. square) under the supervision of the Building Department officials. Care was taken to secure only average workmanship and materials in the construction of the floors.

Test-kilns.—These were made about 11 ft. by 14 ft. in size, inside measurement, with brick walls 12 ins. thick, reinforced by buttresses and iron stays. See Fig. 24. The kilns were 10 ft. high from the upper or main grate-bars to the floor system to be tested, which formed the roof of the kiln. Secondary or lower grate-bars were placed from 14 to 18 ins. below the main grate, air being admitted by openings in the walls below the lower grate. At each corner of the kilns chimneys 15 ins. square were constructed. The floor samples to be tested were constructed between steel beams resting on the brick walls, a different kiln being used for each test. All ceiling surfaces or under sides of floor systems were plastered. Wooden sleepers, with concrete or cinder filling between, were laid over the floor-arches in every case, but no finished wood flooring was used. Care was taken, as before stated, to secure only average samples, such as would be used in ordinary building construction, and in some in-

stances finished samples had to be replaced by the manufacturer on account of more than ordinary refinement in the
work.

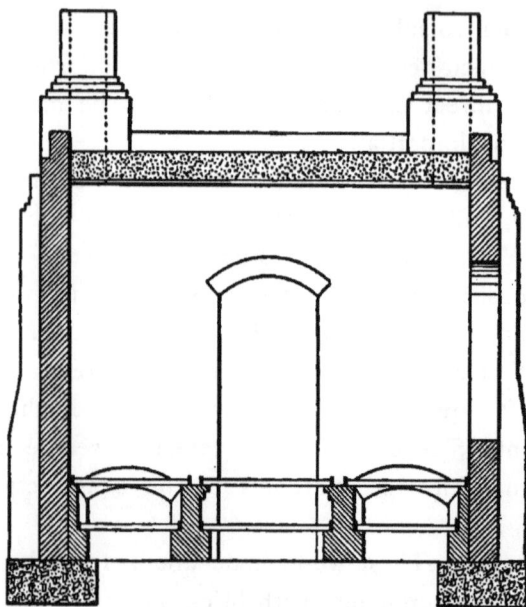

FIG. 24.—Test-kilns used in Tests by New York Building Department.

Method of Testing.—The central panel of the floor system was loaded uniformly to 150 lbs. per sq. ft.   A wood
fire was then started on the grates and kept burning for five
hours, the temperature during the last four hours being kept
as nearly as possible to 2,000 degrees F.;   water was then
applied through a 1⅛-in. nozzle under 60 lbs. pressure, by
the officials of the Fire Department.   This lasted for fifteen
minutes, the first five minutes being on the ceiling only, and
the remaining ten minutes on both ceilings and walls.   The
top of the floor was next flooded with water under low pressure for a space of five miuutes.   At the conclusion of the
fire- and water-test, the original load was removed, and a
similarly placed load of 600 lbs. per sq. ft. was substituted

and maintained for forty-eight hours. This final load was .so placed as to rest entirely on the floor-arch, and not over the supporting beams.

The temperatures were taken by means of the Uehling, Steinbart & Co.'s pneumatic pyrometers, placed in the kiln just below the floor system; also by placing various metals with known melting points at the same positions. Transit observations were taken to determine the combined deflections of both the floor-beams and the arches between them.

The various systems tested, with brief description of results, are as follows.

Rapp Floor.—The supporting 10-in. beams were spaced 4 ft. centres (see Fig. 73, Chapter IX). The initial loading caused a centre deflection after forty-two hours of .075 in. Duration of fire-test, 10.00 A.M. to 3.04 P.M. Maximum temperature recorded, 1,975 degrees.

At the conclusion of the fire-test, six or seven tee-bars had sagged about one inch, but none had dropped from place. All plaster had dropped off except in the corners of the kiln and in the wire lathing around the beam-flanges, but the under side of floor seems to be intact. The maximum deflection was 2.375 ins.

After the water-test, it was found that about 17 per cent. of the bricks had broken or fallen, and would have to be replaced if the floor were to be repaired. Some of the small tees had sagged considerably, the plaster had mostly fallen, and, in places, the wire netting around the beam-flanges had been burned out.

The final test-load on the central arch caused a deflection of .31 in.

Roebling Floor.—The supporting 10-in. 25-lb. I-beams were spaced 3 ft. 10¾ ins. centres. The initial loading of 150 lbs. per sq. ft. caused a centre deflection of .135 in.

Duration of fire-test, 10.15 A.M. to 3.04 P.M. Maximum temperature recorded, 2,300 degrees.

At the conclusion of the fire-test all the plaster was found to be down. The steel beam-flanges and the concrete were red-hot, and even after the application of water, red spots still showed in the concrete.

After the completion of the water-test, the wire netting in the arches had been burned off from both the unprotected arches and those which had been protected by ceiling construction, except in the extreme corners of the kiln. The concrete arches themselves were in good condition, except that some small patches had been washed away in places,

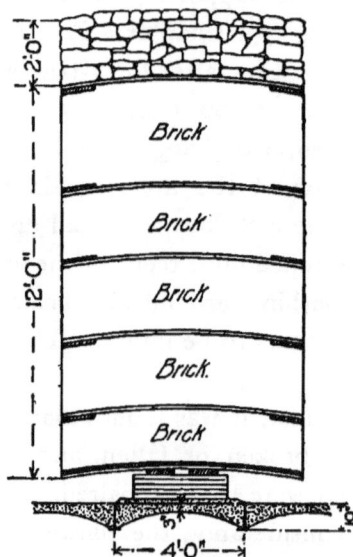

FIG. 25.—Test-load on Roebling Arch.

where the water had been applied with most force. The final test-load of 600 lbs. per sq. ft. caused an immediate centre deflection of .52 in.

In the above test, two partitions of plaster on wire netting, supported by a light steel framework, were erected in the

kiln, extending from grate to ceiling, and projecting about
3 ft. at right angles to the brick walls. These partitions were
completely destroyed.

Roebling Arch—Second Test.—The test-floor just described was subjected to a second test for dead-load only.
The system was used as left by the previous fire- and water-
test, except that a section of the floor 4 ft. long was isolated
from the rest of the floor by cutting openings in the concrete from beam to beam. Timber supports were placed
under the beams and a brick pier 4 ft. by 2 ft. 6 ins. and 1
ft. high was built, as shown in Fig. 25, to receive the load.
This pier was then loaded, as shown, with brick and stone
to the amount of 40,000 lbs. The pile then became too
top-heavy for further additions.

The deflection of the arch under this load was one inch.
After the load was removed this deflection decreased to
$\frac{5}{8}$ in.

Thomson Floor.—This system consists of a flat concrete arch with I-beams only as metal supporting members,

FIG. 26.—Detail of Thomson Floor.

as shown in Fig. 26. The I-beams as used were spaced 3 ft.
6 ins. centres, and were 6-in. 15 lbs. per foot in weight.

The concrete used was made of pea-coal, sand, and Portland cement, mixed with a solution of water and chemicals.
The concrete filling between the nailing-strips was composed
of cinders, sand, and cement. Below the beam-flanges were
placed layers of a mixture of clay, ground feldspar, and
cement, the first layers, nearest the beams, being held in

place by wire netting bent around the beam-flanges. Three
different kinds of plaster were used in different ceiling-bays.
The arch showed no deflection under the initial loading of 150
lbs. per sq. ft. Duration of fire-test, 10.00 A.M. to 3.00 P.M.
Maximum temperature recorded, 2,250 degrees. Maximum
deflection 1.77 inches.

After the completion of the fire-test some plaster had
fallen, but in the centre bay the plastering seemed intact.
The greatest injury to the concrete arch was done by the
water-streams, which caused the falling away of the concrete
around several of the tie-rods and exposed the lower flanges
of the beams in places. Much of the plaster, where not
reached by the water, was still in place.

The final test-load of the 600 lbs. per sq. ft. caused a deflection of 0.38 in. after 42 hours.

M'Cabe Floor.—This arch was made of patent fireproof
blocks supported by tee-irons which rested on the lower
flanges of I-beams spaced 4 ft. centres (see Fig. 27). The

FIG. 27.—Detail of M'Cabe Arch.

arch had three blocks in each space between the beams and
tees, the blocks projecting 3 ins. below the beam-flanges.
The ceiling was plastered with a coat of white plaster.

In this test the brick walls of the kiln were strengthened
by a steel framework consisting of four corner I-beam posts,
horizontal tie-beams, and angle knee-braces at the corners.
This was to give a support for the floor system entirely independent of the brickwork. The supporting floor-beams were
four 10.-in. 33-lb. I-beams. The initial loading caused a deflection of .02 in. Duration of fire-test, 10.00 A.M. to 3.00

P.M. Maximum temperature, 2,325 degrees. Maximum deflection, 0.345 in.

At the end of the fire-test and before the application of the water the ceiling was intact. After the test by water it was found that what would naturally be considered the weakest places, viz., beneath the flanges of the beams and tees, had suffered the most. Five blocks showed open spaces along the lines of the tees, exposing the concrete above.

The final test-load developed a centre deflection of 0.295 in. after 48 hours.

Columbian Floor.—Rolled-steel bars 2 ins. by $\frac{1}{4}$ in., 20 ins. centres, were employed, hung in stirrups from four 10-in. 25-lb. I-beams, 4 ft. centres.

The arch was $5\frac{3}{8}$ ins. thick at the centre, including $\frac{3}{8}$ in. of plaster, being composed of concrete of 1 part Dykerhoff Portland cement, $2\frac{1}{2}$ parts sea sand, and five parts broken blue stone. The lower flanges of the beams were protected by blocks or slabs $1\frac{3}{8}$ ins. thick, made of cement, sand, and broken stone, with hoop-iron embedded therein for clamping to the beam-flanges.

The initial loading showed no deflection. Duration of fire-test, 10.30 A. M. to 3.30 P. M. Maximum temperature recorded, 2,200 degrees. Maximum deflection, at conclusion of water-test, 4.31 ins., the beams having undergone a considerable permanent deflection.

During the fire-test cracks developed between the wooden sleepers and the concrete, and smoke was noticed on the floor surface coming from the sleepers, indicating a charring on the under side. The concrete had sunk very perceptibly in places. After the application of water, the rolled-steel bars originally protected on the under side by $\frac{3}{4}$ in. of concrete, were found exposed to view for about six-sevenths of the entire surface. Along one side of the kiln, the plaster and concrete had remained. Most of the slab-casings under the beam-flanges had been entirely destroyed.   In some cases the iron bands remained attached to the beam-flanges after the concrete which they held in place had been burned or washed away.   The main concrete, however, was intact, showing no holes or openings.

The final load produced a deflection of 0.54 in. after 48 hours.

Bailey Floor.—This floor is made by laying sheets of dovetail-formed metal on the lower flanges of I-beams, and filling the space up to the tops of the beams with cinder concrete (see Fig. 28).   The supporting I-beams used were

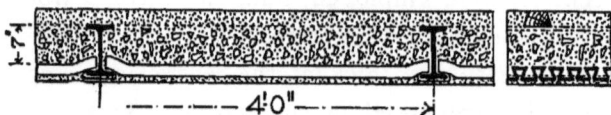

FIG. 28.—Detail of Bailey Floor.

7-in. 20 lbs., spaced 4 feet centres.   The flooring was made of No. 28 gauge sheet-iron, bent to the form shown.   The sheets, after bending, were about 11 ins. wide.   The concrete filling was made of 1 part Rosendale cement, 2 parts gravel, and 4 parts cinders.   The soffit below the metal sheets was plastered with a mixture of plaster of Paris, asbestos fibre, and lime-mortar, strengthened at the lower beam-flanges by wire cloth.

The loading of the 150 lbs. per sq. ft. produced a centre deflection of 0.08 in. after $19\frac{3}{4}$ hours.   Duration of fire-test, 10.00 A.M. to 3.00 P.M.   Maximum temperature, 2,450 degrees.   Maximum deflection, 3.06 ins.

During the fire-test the top of the floor was so hot that it was barely possible to hold the hand upon it.   The ceiling showed good results, with only the white coat of plaster peeled off in places.   Under the action of the water, the plaster was almost entirely washed away, leaving the beams and metal sheets fully exposed.   The top of the floor was not damaged, except by the deflection. The final load caused a centre deflection of 0.695 in., after 48 hours.

Clinton Wire-cloth System.—Segmental wire-cloth centers are used in this system, sprung from, and clamped to the lower flanges of the I-beams. The space is then filled up level to the tops of the beams with cinder concrete, tie-rods being used between the I's.

The supporting-beams were 10-in. 25 lbs., spaced 4 ft. centres. The concrete filling was made of 1 part King's Windsor cement by measure, 2 parts plaster of Paris, and 9 parts cinders. 3-in. by 4-in. sleepers were laid over the I's, filled in between with cinder concrete. The soffit was plastered to a thickness of about $\frac{1}{4}$ in., reinforced under the beam-flanges by wire-cloth nailed to the concrete arch.

The first loading gave a deflection of 0.02 in. after 18 hours. Duration of fire-test, 10.20 A.M. to 3.20 P.M. Maximum temperature, 2,250 degrees. Maximum deflection, 2.565 ins.

During the fire, a large part of the plastering and some of the beam protection had dropped off. All of the balance was destroyed where within reach of the stream of water. Parts of the central arch were found washed away to within about 2 ins. of the tie-rods, but out of the range of the water the flooring seemed uninjured, except that the plastering was gone completely, or was hanging loose. The final load developed a centre deflection of 0.26 in.

Manhattan System. (See Fig. 29.)—The supporting I-beams were 9-in. 21 lbs., spaced 4 ft. centres. A flat centering of wood construction was first placed one inch below the beams. No. 10 expanded metal was then laid on the lower flanges of the I's, the sheets being supported by $\frac{3}{4}$-in. by $\frac{1}{8}$-in. iron hooks, which passed under small iron rods placed under the sheets of expanded metal and over the upper flanges of the I's. Directly on the wooden centers

and to a depth of 4 ins., concrete was placed, well rammed
under and around the beam-flanges, and completely embedding the expanded metal. This concrete was made of 1
part imported Germania cement, 2 parts fine gravel, and 5

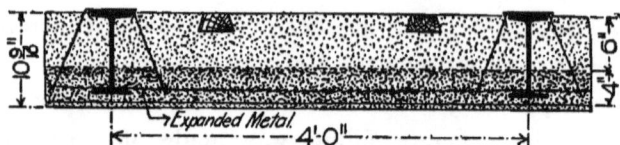

FIG. 29.—Detail of Manhattan Floor.

parts boiler-ash. The remaining 6 ins., up to the tops of
the I's, were filled with a concrete composed of 1 part Globe
cement, imported Belgian, to 10 parts boiler-ash. The floor
sleepers, 2-in. by 3-in., spaced 17 ins. centres, were embedded
in this upper course of concrete. The ceiling was plastered
with King's Windsor cement. The initial loading gave a
centre deflection of 0.02 in.

Duration of fire-test, 10.00 A.M. to 3.00 P.M. Maximum
temperature, 2,150 degrees. Maximum deflection, 2.24 ins.

At the conclusion of the test it was found that the force
of the water had, as is usual in such tests, dislodged the
plastered ceiling where the water struck; while the concrete
in the central bay, which received the greatest force, was
washed out to a depth of one inch above the expanded
metal. Almost all the beam-flanges were completely exposed,
but in the corners of the kiln, where the stream did not reach,
even the lower coat of white plaster appeared unharmed,
indicating that the damage was due to the water and not to
the fire.

The final load of 600 lbs. per sq. ft. caused a centre
deflection of 0.47 in. after 48 hours.

Expanded Metal Co.'s Floor.—The detail of the floor
system used in this test is shown in Fig. 30. The supporting I-beams were 10-in. 30 lbs., 4 ft. centres, tied by ¾-in.

tie-rods. A wooden centering was first placed in position, fitting up between the beams, but leaving a proper outline for the haunches around the beams. Concrete made of 1 part Atlas cement, 2 parts fine sand, and 4 parts cinders, was packed into the haunches and tamped around the beams, and was also spread on the flat portion of the centering.

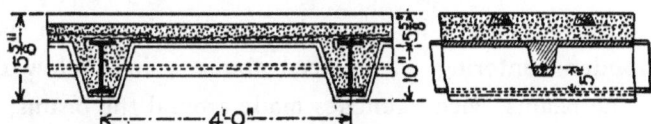

FIG. 30.—Detail of Expanded Metal Co.'s Floor.

Sheets of expanded metal were next laid, one end being bent around a beam-flange, while the other end lapped over and rested on the sheet in the next bay. A second layer of concrete was then filled in, over the expanded metal, to a depth of 3 ins. above the tops of the beams. Sleepers, 3 ins. by 4 ins., were then laid with concrete filling between. The tie-rods, coming $4\frac{3}{4}$ ins. below the main ceiling line, were encased in King's Windsor asbestos cement, as shown in the illustration. The ceiling was plastered with the same cement.

The loading of 150 lbs. per sq. ft. produced a centre deflection of 0.02 in. Duration of fire-test, 10.05 A.M. to 3.05 P.M. Maximum temperature, 2,200 degrees. Maximum deflection, 3 ins.

During the fire the top of the floor system was too hot to touch. The effect of the water was to wash off the plaster and portions of the concrete slab, exposing the expanded metal to view wherever the water had struck. Only a small portion of the plaster around the tie-rods remained, and the lower flanges of the beams were laid bare. In places, the metal lath had fallen away from the beam-flanges. The plaster in the corners of the kiln, away from the fire, appeared intact.   No cracks or bulges appeared in the floor as due to the load above.

The final loading of 600 lbs. per sq. ft. developed a centre deflection of 0.34 in.

Metropolitan System.—The type of Metropolitan floor used in this test was practically that shown in Fig. 72, Chapter IX.   The supporting I-beams were 10-in. 25 lbs., placed 4 ft. centres.   Two $\frac{3}{4}$-in. tie-rods were used in each bay. The wooden centering was placed about 4 ins. below the tops of the beams, with haunches made around the beams, so that the I's should be covered with about 2 ins. of the same filling as used in the arch-slab.   Galvanized-iron cables were used, each made of two wires $\frac{3}{32}$-in. diam., spaced 2 ins. centres.   The lower flanges of the beams were covered with wire cloth.   The filling was made of a composition of plaster of Paris, wood chips, cocoanut fibre, and asbestos.   The ceiling was made of wire mesh fastened to 1-in. by $1\frac{1}{4}$-in. ceiling-bars, 16 ins. centres, hung from the $\frac{3}{4}$-in. round bars in the centre of the span.   The ceiling was plastered two coats.

The initial loading produced a centre deflection of 0.04 of an inch.   Duration of fire-test, 10.22 A.M. to $3.22\frac{1}{2}$ P.M. Maximum temperature, 2,050 degrees.   Maximum deflection, 0.36 in.

At the end of the test it was found that in such places as the water had struck with the greatest force, the arch material was washed away sufficiently to expose the wire cables.   In places out of reach of the water, the arch material was quite soft, and capable of being easily penetrated to a depth of 1 or $1\frac{1}{4}$ ins.   The ceiling was completely gone in places, while other portions had sagged several inches below the original position.   A piece of the composition material, on being removed from under a beam-flange and examined, showed that in the lower or exposed portion all combustible material in the mixture, such as the wood chips and cocoanut fibre, had been completely burned; in the interior portion of the sample such material was charred, while in the upper portion, which came in contact with the beam-flange, the composition was intact.

The final loading caused a centre deflection of 0.26 of an inch after 48 hours.

Fawcett System.—The I-beams used were 6-in. 11.9 lbs., spaced 2 ft. centres. The Fawcett special hollow lintels of terra-cotta were used, resting on the lower flanges of the beams, as shown in Fig. 76, Chapter IX. Cinder concrete made of one part Atlas Portland cement to four parts furnace-cinders was filled in over the blocks to two inches above the I-beams. The floor was finished with a one-inch layer of cement-mortar. The ceiling soffit received a coat of brown plaster and a finished coat of hard plaster.

Duration of fire-test, 9.33 A.M. to 12.16 P.M. Maximum temperature 2,200 degrees. Maximum deflection, 2.125 ins.

During the progress of the fire, large cracks developed in the upper part of the floor, and most of the plastering on the ceiling fell off. The application of water on the hot tiles caused a great deal of damage. The greater portion of the tiles cracked, allowing pieces to fall, and in many places the concrete had been so exposed that it was washed away to some depth.

The iron beams sagged badly in the centre, one beam having a permanent set of $5\frac{1}{2}$ ins. The deflection due to the final load alone, without considering the permanent set, was 0.87 of an inch.

Porous Terra-cotta Arch.—This floor was furnished by the Central Fireproofing Company. The floor-beams used were 10-in., $23\frac{1}{2}$ lbs., spaced 4 ft. centres, with two $\frac{3}{4}$-in. tie-rods in each bay. The arches were made of 10-in. end-construction porous terra-cotta blocks, levelled up to the tops of the beams with a one-inch filling of cement-mortar. Nailing-strips, 2-in. by 4-in., 16 ins. centres, were then laid, filled in between with cinder concrete made of one part cement to three parts cinders. The beam-flanges were protected by terra-cotta slabs held in position by bevelled lips on the skew-backs. Two of the bays were plastered on the soffit, while one bay was left with the tile exposed.

Duration of fire-test, 9.00 A.M. to 3.00 P.M. At 2.54 P.M. cast-iron was placed in the kiln and melted in two minutes. Maximum deflection, 2.16 ins.

During the fire-test all of the plaster had dropped from the walls of the kiln. Upon the application of water all of the plaster on the ceiling fell. About 35 per cent. of the arch-blocks were found to be cracked, the lower sections of some blocks having broken off to a depth of about $3\frac{1}{2}$ ins. One block was hanging half out of the arch. All of the soffit protection-pieces had fallen except very near the walls of the kiln.

The final load of 600 lbs. per sq. ft. gave a deflection 0.22 in.

Porous Terra-cotta Arch. Second Test.—Twenty days after the final loading of the arch described above, the supporting beams were shored up to prevent deflection and the central bay of the floor was loaded up to 611 lbs. per sq. ft. On following days this load was gradually increased to 1,175 lbs. per sq. ft. The deflection due to this load was 0.84 in. The load was then shifted to cover an area of 9 ft. by 4 ft., the total load being 1,960 lbs. per sq. ft. The deflection was 2.2 ins. This, added to the permanent set which existed previous to loading, gave a total deflection of 3.41 ins.

The arch was still intact under this load.

Guastavino Floor Construction.—The brick walls of the kiln in this case were corbelled out 6 ins., and a rectangular horizontal iron frame was built into the upper portion of the

kiln-walls to take the thrust of the dome. This system required no intermediate floor-beams. The dome consisted of
three successive courses of flat fire-clay tiles, which sprung
from the side walls with a rise at the centre of 10 per cent.
of the greatest span. The tiles were 6 ins. wide, 12 ins. to
18 ins. long, and ¾ in. to 1 in. in thickness, being laid in
cement-mortar. The construction was built on a wooden
center. Two kinds of floor surfacing were used. In one
half,, brick ribs extended over the dome, supporting a double
course of horizontal flat tiles, which received the finished floor.
In the other half, cinder-concrete filling was used, with embedded nailing strips. Portions of the ceiling-soffit were
plastered, and some spaces of tile were left exposed.

The initial loading gave a centre deflection of .017 in.
Duration of fire-test, 9.15 A.M. to 3.18 P.M. Maximum
temperature 2,525 degrees. Maximum deflection, 0.71 in.

Nearly all of the plaster fell early in the fire-test, but before
the application of water no cracks had developed in the ceiling,
and no tiles had fallen. During the water-test, portions of
the lower course of tile fell in pieces, due to the sudden contraction. One I-beam, supporting a corner smoke-flue, became exposed. It was noticed that the centre of the dome
rose gradually under the influence of the applied heat, which
caused the expansion of the masonry in the dome construction. The greatest elevation was 0.71 in.

The final loading of 600 lbs. per sq. ft. gave a centre deflection of 0.195 in.

Hard-burned Terra-cotta Arch. — This floor was furnished by the Metropolitan Fireproofing Company. The I-
beams were 10-in. 25-lbs., 4-ft. centres, with two ¾-in. tie-
rods in each bay. The terra-cotta blocks were 10 ins. deep,
side-construction, projecting 1¼ ins. below the I-beams.
They were furnished by the Raritan Hollow and Porous Brick
Company. Each arch-ring consisted of two skew-backs, four

voussoirs, and one key-block. The skew-backs protected the beam-flanges, without the aid of any separate pieces. The blocks were laid in cement-mortar, ¼-in. joints. Bevelled sleepers 2 ins. by 4 ins., 18 ins. centres, were laid over the arches with cinder-concrete filling. The ceiling was plastered two coats.

The loading of 150 lbs. per sq. ft. gave no deflection. Duration of fire-test, 10.22 A.M. to 3.22 P.M. Maximum temperature 2,050 degrees. Maximum deflection, 1.84 in.

At the conclusion of the test about all of the plaster was gone, even where not reached by the water. A considerable portion of the tile was broken by the effects of the water, causing the lower parts to fall, and thus exposing the inside sections of the tile. Many of the skew-backs had broken directly under the beam-flanges, leaving the latter partly exposed.

The final load of 600 lbs. per sq. ft. caused a centre deflection of 0.22 in.

Fire-tests of Wooden Floor Construction. — These tests, in addition to the fireproof floor-tests just described, were made by the New York Building Department. The furnace used was designed by Mr. Stevenson Constable, who had charge of the other tests. It consisted of a box of sheet-iron about 9 ft. long, 5 ft. wide at the top, tapering down to 3 ft. wide at the bottom, and 2½ ft. deep. It was lined with 9 ins. of fire-brick. A section of the flooring to be tested was used as a cover at about the middle of the top of the trough or furnace. A section 4 ft. square was used. Fire was supplied by two jets of crude oil through steam jet oil-burners, with steam at 50 lbs. pressure.

Test No. 1.—Section consisted of two thicknesses of ⅞-in. tongued and grooved yellow-pine flooring, with a sheet of asbestos felt placed between. A maximum temperature of 1,560° was attained. Smoke first appeared at the joints 18 minutes after start. Flame first appeared 29 minutes after the start.

Test No. 2.—Same section as in Test No. 1, except that a layer of "Salamander" was used between the floorings instead of the asbestos. The first smoke appeared 27 minutes after the start, and flame was first seen 34 minutes after the start.

These two tests indicate the superiority of "Salamander" lining over asbestos, as the former material required 40 minutes to reduce the sample to the same condition reached by the latter in 29 minutes.

Test No. 3.—Was a Mill Construction floor, made of a lower course of 2-in. tongued and grooved spruce planks, covered with an upper course of $\frac{7}{8}$-in. yellow-pine. No fireproof filling was used. The first smoke appeared in $63\frac{1}{2}$ minutes from start, and the first flame was noticeable in $67\frac{1}{2}$ minutes.

Test No. 4.—Same flooring as in test No. 3, except that a single layer of "Salamander" was laid between the courses. The first smoke appeared in 62 minutes, and the first flame was visible in 1 hour and 35 minutes from the start of test.

Tests Nos. 3 and 4 showed that the lack of protection at the joints constitutes the main weakness for fire-resisting qualities in Mill Construction floors.

CHAPTER V.

MATERIALS USED IN FIRE-RESISTING CONSTRUCTION.

FIRE-RESISTING construction depends largely upon the efficiency and permanency, under fire- and water-tests, of the materials employed either for the essential structural portions of the building, or for the insulating or protective coverings which may be applied to the load-bearing members. Such efficiency and permanency are influenced by the materials themselves, and also by the quantity, form, and method of application employed.

But even the best materials, used in the most discriminating manner, are limited as to their effectiveness and endurance by the amount of time and exposure; and unless the intensity of fire and the time during which it is applied can be limited so as to secure a reasonable factor of safety for the materials employed, the possibility of constructing a really fireproof building is frustrated. For instance, brick masonry, than which no better fire-resisting material is known, may be seriously damaged by a long-continued fire of severe intensity.

Notwithstanding these limitations, past experience certainly justifies the statement that steel-frame buildings may be successfully and economically designed so as to render them practically fireproof in themselves. Some damage may be done to the structure, and the contents may be destroyed either in part or entire; but if the portions of a building which is subjected to fire can so withstand the effects of heat and applications of cold water as practically to retain their normal form and position, they may be said to possess the qualities of fire-resistance; for if the important structural portions of the building can retain their position and practical integrity under exposure, the spread of fire will be prevented, and the required service of such parts of the structure will have been fulfilled, even though a partial reconstruction of these exposed portions becomes necessary. Those materials which would require the least reconstruction or repair after service in fire, should be selected.

The fire-resisting qualities of materials as ordinarily used will here alone be considered. The equally important questions of general design, detail, and equipment, which will be found to aid largely in minimizing the damage to the structure itself as well as to the contents, will be treated of elsewhere.

The materials usually employed in fireproof building construction are:

1. Cast-and wrought-iron and steel.
2. Stone.
3. Brick.
4. Terra-cotta.
5. Mortars, plasters, plaster of Paris.
6. Concrete.

Cast- and Wrought-Iron and Steel.—As the result of tests and actual experience in conflagrations it may be stated that unprotected cast-iron can stand practically unharmed up to temperatures of 1,300 or 1,500 degrees F. while carrying very heavy loads, even with frequent applications of cold water while the metal is at a red heat.

Various fires show that cast-iron columns will stand and safely carry their loads under higher temperatures than are commonly supposed. In the Ames Building fire in Boston in 1889, a majority of the unprotected cast-iron columns were found to be in a sufficiently good condition to warrant their re-use. The same experience has been found in English and American cotton mills where cast-iron columns have been used a second time. Photographs taken of the Horne Buildings in Pittsburg, which were destroyed by fire (see Chapter III), show the cast-iron columns in the first story at the front of the Office Building to be apparently uninjured, though subjected to a very severe heat.

Wrought-iron or steel will commence to yield at temperatures of from 1,000 to 1,200 deg. F.

Very few tests of recent date have been made to determine the action of iron and steel members while carrying loads under fire and water exposure. One of the most recent, as well as one of the most valuable series of tests ever made on cast-iron and steel under high temperatures, was conducted in New York City in 1896 by a committee representing the Tariff Association of New York, the Architectural League of New York, and the American Society of Mechanical Engineers.* The tests were few in number, but most important in results.

Five full-sized columns were tested in brick furnaces which were built for the purpose. The columns were made of forms and lengths as representing common practice, and they were placed in compression by means of a hydraulic ram to obtain loadings approximating those found under ordinary conditions.

Test No. 1. The column tested was unprotected, box-shaped, made of two steel channels and side plates. The highest temperature recorded was 1,230 deg. F. After an exposure of 1 hour and 21 minutes the column began to yield under a load of 46.00 tons, the temperature being 1,210 deg. F. The column buckled at the centre by the

---

* See *Engineering News*, August 6, 1896.

wrinkling of the plates. The breaking-load was computed by Gordon's formula to be 342 tons.

Test No. 2 consisted of an unprotected 8-in. standard steel Z column. The maximum temperature recorded was 1,375 deg. F. A uniform loading of 84.8 tons was maintained during the entire test. The column commenced to yield after an exposure of 24 minutes, the temperature being 1,125 deg. Deflection occurred at the lower third point. The computed breaking-load was 303 tons.

Test No. 3 was of a cast-iron, round, hollow column, with faced flanges at both ends. The highest temperature registered was 1,250 deg. F. Deflection at the centre occurred in 1 hour and 8 minutes after start of test. The load was 84.8 tons, temperature 1,137 deg. The computed breaking-load was 451 tons, safe load 90.2 tons.

Test No. 4 consisted of an unprotected cast-iron, round, hollow column. Bending, at about the centre of the column, started after 35 minutes of exposure, under a loading of 84.8 tons and a temperature of 1,350 deg. Eight minutes later, under the same loading and a temperature of 1,550 deg., fracture occurred at the centre of the column where the deflection was greatest. A crack was also developed above the point of fracture on the convex side.

Test No. 5 combined a fire- and water-test on a cast-iron column, 8 ins. diam. by 1-in. metal. The maximum temperature recorded was 1,300 deg. F. The column started bending in 2 hours and 15 minutes after the beginning of the test, after several applications of cold water. The temperature was 1,275 deg., load on column 84.8 tons. At the conclusion of the test the column was found to be badly bent, but was otherwise uninjured, although the column was at a red-heat when water was last applied.

The results of these tests should not serve to detract from the importance of adequate fireproofing. They point, rather, to the absolute necessity of such protection. A building in the compact mercantile portion of a city should be designed to withstand a temperature arising from the possible consuming of its contents, and such conditions would probably generate a heat considerably greater than 2,000 degrees. The heat of a wood fire is from 800 to 1,140 degrees; charcoal, about 2,200 degrees; coal, about 2,400 degrees. It has been shown that failure of cast-iron or steel columns would begin at from 1,000 to 1,500 degrees, hence the necessity for a proper protection from fire is emphasized, and the importance increases in proportion to the load to be borne.

These remarks are equally applicable to all other load-bearing members, whether of cast- or wrought-iron or steel.

Expansion. — Unprotected or poorly protected steel frames will, under the action of high temperatures, so expand as to cause the deformation of the structure, if not complete ruin. For each degree F. of elevation of temperature soft steel or iron will extend about $\frac{1}{150000}$ part of its length. For each 100 degrees F. increase in temperature the increase in length would be about one inch in 125 feet. Where unprotected iron or steel beams or girders are supported by masonry walls, this expansion is sufficient to cause the overthrow of the bearing walls. Where skeleton or cage construction is employed, the protection must be sufficient to prevent all expansion or distortion.

The Engineers' Report on the Pittsburg fire contained the following references to this consideration :

" In buildings of about this height [6 stories] the distortion of the steel framework due to the heat of the fire cannot in any instance be sufficient to work any serious damage ; nor is it probable that at any time would connection rivets be sheared off. This conclusion is arrived at for the reason that there is no probability that any future fire will be fiercer than the one at issue.

" The column protection, although composed of the very best obtainable kind of fire-clay tile, was also not of sufficient strength.

" In our opinion it would have been necessary to dismantle the whole steel framework had this structure been fourteen or fifteen stories high. The leaning at that height at the same proportion as developed, would have entailed the necessity of taking the whole structure down."

Effects of Water under High Temperatures.—"The effect of quenching with water, either iron or steel when red-hot, depends entirely upon the form of the test-piece, and upon whether it is under the action of external stresses. If the piece is perfectly homogeneous and symmetrical, and heated and quenched uniformly, nothing will happen beyond the alteration of the temper of the steel and the possible chilling of the surface of the iron; if the metal is not homogeneous and the cast-iron be at all spongy in parts, the quenching will most probably cause it to fly into pieces, principally on account of the steam generated in the crevices. But if it forms part of a structure the conditions are entirely different. If a simple riveted plate-girder without any load is heated and cooled, either suddenly or otherwise, it will not return to its original form, on account of the internal stresses set up in the various operations during construction. If the girder or column is loaded unequally, or is in any way constrained, it will neither expand uniformly with red-heat nor return to its original form or position when cooling. The extent of the distortions depends, of course, entirely upon the conditions referred to; beams and columns of wrought-iron and steel may be twisted out of all shape, and if of cast-iron may be fractured with cooling." *

Stone.—It is very common after severe fires to see stone fronts burned away or destroyed, leaving the brick backing

---

* See John J. Webster in Proc. Inst. C.E., Vol. cv, 1891.

standing for several stories in height.   Stone, under the action of severe heat, will crack, shell, or calcine, according to the nature of the material.   It should be used most cautiously in fireproof construction.

Granite will explode and fly off in fragments, or it will disintegrate into a fine sand.   In some building laws the non-fireproof character of granite is clearly recognized, in that brick or terra-cotta protection is required for granite support-ing-members.   The face of granite stones will spall, or split off, and this often with considerable explosive violence. Granite mullions which have been exposed to flame may commonly be seen in which the exterior corners have so split off as to leave the face V-shaped.

Limestones and marbles are damaged by heat more than any other building stones.   They become calcined under in-tense heat, or decomposed into lime.   This has been clearly demonstrated in many fires.   Limestone fronts have been totally destroyed, while the brick backing has often remained comparatively uninjured.

Compact, fine-grained sandstones withstand the action of fire better than any other stone usually employed.   In some cases, where exposed to moderate heat, sandstones have been uninjured except for discoloration due to smoke, but in the Boston fire on Bedford Street, in which the brown sandstone buildings designed by H. H. Richardson were destroyed, the sandstone was badly affected by the fire and water.   The stone will flake off in considerable quantities, and its destruction is simply a question of time.

In general it may be said that no building stone will with-stand great heat as well as does brickwork.

A comparison of the action of building stones and terra-cotta under fire- and water-tests is described by Thomas Cusack in *The Brickbuilder* for January, 1899.   Three-inch cubes of sandstone, granite, limestone, and marble, and an equal number of terra-cotta blocks, representing the same color equivalents, were tested. The stones were heated for ten minutes, and were then allowed to cool gradually. The terra-cotta blocks were heated for fifteen minutes to nearly white heat, and were then placed in cold water. The lime-stone and marble were largely reduced to powder; the granite began to fuse at a rose-red heat, and when cool fell asunder; the brown sandstone showed the greatest resistance, but on cooling, cracked and became worthless. The cube of terra-cotta corresponding to the sandstone sample showed a fine crack along the face, but it was still capable of withstanding a crushing-weight of 2,500 lbs. per sq. inch. The other samples of the terra-cotta underwent little change, in spite of the sudden reaction from white heat to cold water.

The appraisers' report of the Pittsburg fire states that "The ashlar and cut-stone on the front and side, where subjected to heat, were a total loss, and in addition to replacing them with new ones, it costs considerable to remove them and cart them to the dump."

The total destruction of the marble front in the Home Life Building was pointed out in Chapter III. Above the eighth floor it was necessary to provide an entirely new façade. In the side-walls, also, many of the stone window-lintels were cracked and splintered.

Further data respecting the use of stone fronts, etc., for buildings will be found in Chapter XII.

Brick.—Many conflagrations have fully demonstrated the fire-resisting qualities of good brickwork. Brick and terra-cotta, both of which are clay products, have stood repeated fire-tests when nearly all other ordinary building materials have been either seriously damaged or totally destroyed.

The ability of brick to withstand fire depends upon the

method of manufacture and upon the chemical properties of the clay employed.   When the old style up-draught kiln was used for the burning of the brick, the position of the brick in the kiln affected the fire-resisting properties.   The clinker or arch bricks, which formed the arches in which the fire was built, were usually overburned or partially vitrified.   These possessed admirable fire-resisting properties, but for use in load-bearing walls or piers were too weak and too brittle, although very hard.   The soft bricks, which formed the exterior of the kiln, were usually underburned and too soft for ordinary use.   The body or hard bricks, in the interior of the mass, could alone be used for the best results under load- and fire-resistance.

With the newer styles of permanent down-draught kilns, the position of the brick during the burning is much less important than was formerly the case, as the quality is very nearly uniform throughout the kiln.   The chemical composition of the clay is now the most important factor in determining the fire-resistance of the brick.

The fire-resisting properties depend chiefly upon the amounts and properties of silica and alumina in the clay, and also upon the amounts of oxide of iron, lime, magnesia, potash, etc.   Common clay, used in the manufacture of common brick, consists principally of silicate of alumina, lime, magnesia, and oxide of iron.   The latter ingredient adds to the hardness and strength of the brick.

"Uncombined silica, if not in excess, is beneficial, as it preserves the form of the brick at high temperatures.   In excess it destroys the cohesion, and renders the bricks brittle and weak.   Twenty-five per cent. of silica is a good proportion." *

For fire-bricks intended to resist extreme heat, without

* See " A Treatise on Masonry Construction," I. O. Baker.

heavy loads, silica should be used in excess of the proportion stated above. "The presence of oxide of iron is very injurious, and, as a rule, the presence of 6 per cent. justifies the rejection of the brick. In specifications it should generally be stipulated that fire-brick should contain less than 6 per cent. of oxide of iron, and less than an aggregate of 3 per cent. of combined lime, soda and potash, and magnesia. The sulphide of iron—pyrites—is even worse in its effect on fire-brick than the substances first named."

Good fire-resisting brick should be of homogeneous composition and texture, regular in shape, uniform in size, strong, and infusible.

In the fires which were described in Chapter III, the endurance of brickwork is especially noticeable in the Pittsburg and Home Insurance Building fires. Even the face-brick was but little injured. In the appraisers' report on the former fire, it is stated that "The brickwork stood the strain remarkably well, and there would not have been 15 per cent. loss on it if the other members of the structure had stood the strain as well." In the Home Insurance Building fire, the side, court, and rear walls of brick were badly smoked, but their good condition was in marked contrast to the destroyed stone front. It was not found necessary to rebuild any great portions of the brick walls, a thorough cleaning and repainting being sufficient.

Structural Terra-cotta.—The terra-cotta used for structural purposes, as for floor-arches, column protection, and for partitions, is either "Porous," "Semi-porous," or "Hard-burned," according to the method of manufacture. Porous terra-cotta is also called terra-cotta lumber, cellular pottery, soft tile, porous tile, etc.; while hard-burned terra-cotta is sometimes called fire-clay tile, hard tile, or dense tile.

Manufacture of Porous Terra-cotta. — Porous terra-cotta is made by mixing sawdust or finely cut straw with pure clay, which is then moulded and burned under a high heat, causing the combustion of the straw or sawdust, and leaving the material in a porous state, thereby reducing the weight of the original mass. The factories or places of manufacture are usually located near an adequate supply of clay of the required properties. The clay is dug from the " clay-bank," taken to the clay-house, where it is broken into pieces as small as practicable by hand labor and mixed with coarse soft-wood sawdust (pine or spruce) or finely cut straw, one volume of sawdust or straw to two volumes of clay. During the wet season this mixture is tempered with a quantity of either dry clay or crushed brick, to prevent unusual shrinkage due to the large volume of water in the clay. The mixture is passed through a disintegrator, consisting of an endless worm or cutter revolving in a sloping trough, which thoroughly cuts up all the clay before conveying it to the "pug-mill." The "pug-mill" consists of a hopper at the top, leading down between a set of two corrugated rolls revolving in different directions. These corrugations crush the clay between them, allowing stones of about one inch diameter or less to pass through whole. Large stones are separated from the clay and are delivered to the refuse-box. A second and lower set of smooth rolls, revolving in the same directions as the first set, crushes the clay and small stones into a thoroughly mixed and tempered state, distributing the sawdust through the mass very evenly. In dry seasons water may be added in required quantity at the hopper to produce a plastic mass.

From the pug-mill base a conveyor receives the clay and carries it to the machine which forms the tile. These conveyors have different forms, but are commonly of either a continuous belting of rubber or a series of slats in the form of a belt.

The "tile-machines," or machines which shape the stream of clay, are of various patterns, each of which have points which commend them, but the natures of the clay, according to locality, require different machines. The machine takes the clay, and by different means again works the mass, tempering it by a set of revolving knives into a sufficiently soft and plastic state to allow the required shaping. In forming the blocks these machines are operated differently, according to pattern of machine used. Some force the stream of clay by means of a hinged cam. This cam is moved forward or backward by another cam fastened to the centre revolving shaft, and coming in contact with the hinged cam, forces the clay before it. Others, in the plunger style, move the clay by means of a piston-head, operated directly from an independent cylinder. In either case the clay is forced from the interior of the machine through the "form," which is made of two independent parts—the "plugs" or dies, which are within the machine—and the "form" proper, which constitutes the outlet.

The plugs are made of metal, and are of the exact shape and relative position of the voids in the tile-block. They are stationary, with their faces placed a few inches inside the final form. As the clay is pressed by the plungers against the faces of the plugs it is perforated, thus forming the voids in proper position for the final blocks. After passing the plugs the clay has no external shape, until, by the continued operation of the plunger, the clay is forced from the form or die placed at the outlet of the machine, thus giving the required exterior shape of the manufactured product.

This finished shape is forced out continuously on to the cutting-table, which is usually composed of sets of rolls or plates, well greased to prevent adhesion and friction, which would have a tendency to deform the block. Above the cutting-table are the cutters, which are of various styles, fastened in many ways, but on the general principle of an "arbor" or light framework which spans the table, the opposite sides being connected by wires or knives. This frame is moved up and down by means of a lever controlled by the operator, who cuts the moving mass of clay into blocks of the required shape and length. The wires may be parallel, as for filler-blocks, or made to cut the form of a key.

The manufacture of special forms of terra-cotta, although not differing widely from the ordinary shapes, compels some special manipulation—as, for instance, in making skew-backs. The side construction skew-back, when run from the forcing-machine, comes out with the shape of the beam already formed, while in the case of the end-construction skew-back, the seat of the skew has to be formed by hand after it has been run from the machine. It is cut to the shape desired over a templet and by means of a wire fastened to a bow. Other products, such as rabbeted ceiling or roofing blocks, have to have the rabbet formed in a like manner by hand where the blocks are over thirteen or fourteen inches in width, because the form or mouth of the machine cannot produce a wider stream of clay. Such blocks can be made complete by the forming-machine when not over the above width, as in this case the voids and rabbets can be parallel; but in wider blocks the voids run at right angles to the rabbets, so that the latter must be hand-formed. For circular column covering-blocks a form of the same shape is usually placed on the cutting-table, onto which the clay is forced when coming from the machine. This style of block is usually dried standing on end to prevent deformation by sagging.

From the cutting-table the blocks are placed upon "pallets," consisting of light open wooden gratings, which, when filled, are stacked upon cars made of light metal framing with adjustable racks to receive the pallets in such manner that the blocks just clear each other and permit the free circulation of air over the entire area of the blocks. The cars are then run into the "dry tunnels," which are heated by means of steam-coils to a temperature of about 200 to 300 degrees for a space of time varying according to the nature of the material and the size of the blocks, usually taking from two to three days. At the end of this time the blocks are sufficiently dry to permit handling.

The blocks are next stacked in kilns, which are of various styles, the "down-draft" pattern being usually considered the most satisfactory. This style of kiln consists of a large brick vault or chamber, with a double floor (the upper one of which is perforated like a grating), double side walls, with a space between, single end walls with a door in each, and a double-vaulted roof. A series of furnaces is placed along each side, at the bottom of the kiln. The heat is conveyed from the furnaces to the roof or dome of the kiln by means of the open space between the inner and outer side walls, thence down into the interior by means of the openings in the interior roof, and so down through the chamber and the perforated floor to the chimneys and out. The temperature is usually from 2,100 to 2,500 degrees, for a time depending on the quantity and quality of the contents,—usually about three or four days of continuous heat.

On starting heat, the first thing done is to "gas" the kiln, the purpose being to remove the surplus moisture slowly, in order to prevent great cracking of the tiles. This is done by applying a slow heat, continuing usually from twelve to twenty-four hours.

After being thoroughly fired in the kiln, the tile is ready for use. The sawdust in the clay is entirely consumed during the firing, leaving the finished product in a finely honeycombed state.

If the clay used is of a granular nature, the combustible

material used to produce the porosity should be of but slight quantity in comparison with the total bulk. If of large quantity, the film which originally encased the sawdust or other combustible material before burning is so light that when burned it leaves the finished product full of large cells. This will give an insufficient strength. If the clay is of a fibrous nature it will take a much larger quantity of sawdust, and a much stronger block, comparatively, will be produced. It is generally conceded that fibrous clay makes a much better porous material than granular clay. These factors should be taken into consideration in determining the texture of porous material.

The manufacture of "semi-porous" terra-cotta differs from that of porous terra-cotta principally in the composition of the mixture. To a fair quality of fire-clay, containing about 60 per cent. of silica, is added a certain percentage of clean calcine fire-clay, coarsely ground, and a percentage of coarsely ground clean bituminous coal. This mixture is thoroughly tempered, and burned to the desired shapes. The result of this mixture is a material slightly more porous than the best grade of fire-brick, and still not as soft as the ordinary porous terra-cotta made with sawdust. It is claimed that semi-porous tile may be heated to a temperature of 3,500 degrees, and immersed in cold water at that temperature, without cracking.

Manufacture of Hard-burned Terra-cotta. — Hard-burned terra-cotta is made of natural clays without the addition of any combustible material. The only ingredients added to the natural clay in making this product are, in low grades of clay, crushed brick or sand, to prevent abnormal shrinkage. During manufacture the clay is subjected to a high pressure which gives the material a dense texture, and great strength under crushing-loads. The blocks are shaped by the forming-machine, as before described, and they are burned in kilns, like the porous product, except that the time required for burning is longer.

If the material is quite rough, it indicates too great a quantity of sand, which produces undesirable brittleness. No dependence can be placed on a test of strength of such material. If clay with an excessive quantity of sand is burned at a low heat, it will not have been sufficiently burned to fuse or unite the particles of sand, thus producing a weak and brittle block. If burned at a high heat, sufficient to fuse the sand, it is nearly, if not quite, vitrified, in which case suction is almost wholly destroyed. Hence a hard, rough material is an undesirable building terra-cotta, because it has been burned either too much or not enough. The product should have a *hard* but *smooth* surface.

Comparison in Fire-resisting Properties.—The tests made in Denver, Colo., in 1890 would seem conclusive as to the relative values of porous and hard-burned terra-cotta under the action of fire and water, but the latter material is still extensively employed, although many fires and tests since 1890 have fully confirmed the results of the Denver tests.

The value of hard-burned terra-cotta as a heat insulator depends entirely upon its cellular structure, the only protection being afforded by the non-heat-conducting air-spaces. The material itself conducts heat much more readily than the porous kind. To be efficient, the air-spaces must be of adequate size and number to insulate the material to be protected.

When cooled by water, sudden contraction is liable to occur, thereby cracking the blocks. If made of a good refractory clay, blocks with two or more air-spaces are very liable to have the outer webs destroyed under this action, as was well illustrated by the hard-tile floor-arches in the Horne Store Building fire, where the ceiling-slabs were largely broken away. This was due to the inability of the material to withstand the inequalities of expansion and contraction caused by the heating of one side of the arches only. The blocks usually break first in the corners, both because the strain is greatest there, and the tile weakest. "The strain is greatest in the corners because the expansion of the one side tends to shear it from the adjoining sides, and it is weakest in the corners because if there is any initial stress in the material, it would more naturally occur there than elsewhere."

Even if not cooled with water, other fires have shown that hard-burned terra-cotta will crack and fall to pieces under severe heat alone. This was demonstrated in the Schiller Theatre fire.

Porous terra-cotta is non-heat-conducting in itself, without reference to its form. It is made in solid as well as in hollow forms, but the cellular structure undoubtedly improves its efficiency. The best products of a porous nature have resisted fire and water far better than the best hard tile. For column and girder protection, where the blocks do not carry loads, the porous material is very generally used, but in floor-construction many architects prefer to use the hard-burned variety on account of its greater strength and cheaper price.

If porous terra-cotta is not burned at a sufficiently high temperature to consume all of the sawdust, the throwing of cold water upon the heated surfaces will cause an expansion or disintegration due to the absorption of the water and its conversion into steam.

Ornamental Terra-cotta.—Ornamental terra-cotta, like bricks and structural porous and hard-burned terra-cotta, is made of clay, but of a much finer and selected quality, pure, and free from stones and grit.

Distinctly different qualities of color pertain to the clays found in different localities, and it is only by the mixture of two or more varieties of clay from different sources that certain color-effects can be obtained. Formulas for these effects of mixture are carefully preserved by the leading terra-cotta companies. A great variety of excellent "tide-water" clays is found in Northern and Central New Jersey, and large quantities are mined annually for the manufacture of both structural and ornamental terra-cotta, as well as for other clay products, such as tiles, pottery, and bricks. The New Jersey clays are of sufficient variety to give, in themselves, almost any color-effect desired. "Mountain clay," containing a much greater quantity of silica, is mined in Virginia, and is used especially for the production of white terra-cotta. The clays mined in Ohio are shale-clays, and are not used for ornamental purposes.

After being brought to the factory from the clay-banks, the material is allowed to remain in bins to temper or season for a length of time depending largely upon experience. When this has been accomplished "grog" or "grit" is added to the clay to cause a partial vitrification of the mass during burning, and to insure an even and not too great shrinkage. This "grog" or "grit" consists of pieces of old ornamental terra-cotta which has been burned and then broken by crushers, old pottery or fire-bricks finely ground, and fine white silicious sand.

This mixture is then taken to the crusher, which consists of two steel rollers working about a vertical shaft upon circular beds. Here the mass is thoroughly ground and crushed until the well-worked material is forced over the edges of the circular beds into the conveyors, where water is added. The mass is next passed through the pug-mill or mixer, where it is cut by knives and thoroughly mixed.

After passing through this process the material has the consistency of putty; it is plastic and easy to work. For convenience in handling it is cut into small cakes for use in the moulding-rooms.

In the preparation of the models and moulds for ornamental terra-cotta, a very high grade of skilled labor is required.   In the modelling-rooms of the terra-cotta companies a large number of sculptors or modellers are employed
to produce the great range of designs called for by the
artistic treatment of the terra-cotta product.   If, as is usual,
several or many pieces are to be of the same pattern, the design is first executed in clay, and submitted to the architect
for approval, if required.   From this clay model a plaster
mould, or negative, is made, often in several pieces, fitted
together in a variety of ways, to allow the "drawing"
of undercut work, or the removal of the mould from a
several-sided piece.   When the moulds have been fitted
and dried sufficiently they are sent to the pressing department.

Here the "pressers" carefully work the plastic clay by
hand into the moulds, where it is allowed to remain for about
a day until partially dry.   The moulds are then removed,
and the clay possesses sufficient stability to allow careful
moving to the drying-rooms.   Some clays, especially in
large pieces, are allowed to dry in the air a week or
more before placing under forced heat.   Before placing in
the drying-room, the blocks are retouched and gone over
with tools to remove imperfections or to add mouldings or
delicate ornamentation.   The drying-rooms are fitted with
steam-coils, and the blocks are kept here on cars for about a
week, at a temperature of about 150 degrees.   When removed, the clay must be "bone-dry."   The pieces are next
placed in the kilns, which are commonly of the down-draft,
"beehive" pattern.   These are considered to be the room-
iest and most convenient, as well as the most satisfactory as
regards the heating-surface, attainable.   The shape of the kiln
resembles a huge beehive, in which the fires are maintained
below the floor.   The heat ascends through the hollow outer

walls, and then returns downward through a central cylindrical shaft.

The pieces are kept in the kiln, under heat, for a day or two, with the kiln doors entirely open. This is to allow the escape of any surplus moisture which the blocks may still retain. The doors are then sealed, and continuous heat to a temperature of at least 2,300 degrees is applied for ten days or two weeks. The clay is brought to a white heat, and, in burning, the alkaline salts contained in the clay produce an efflorescence which, acting upon the silicates of the surface, produces a vitrified skin or coating over the mass. Samples are drawn from the kiln from time to time, through small openings provided for the purpose, until the necessary condition known by experience is indicated.

Very light shades of terra-cotta are obtained by spraying a coat of liquid clay of the required properties over the blocks of darker body, before drying. This clay surface unites with the body, by fusing, so thoroughly that no joint remains between the two clays, and peeling is impossible.

The chief recommendations of ornamental glazed terra-cotta lie in its durability and its admirable fire-resistance. Although of comparatively recent introduction to building methods in this country, the use of this material has proved eminently satisfactory and it is now considered the most acceptable material for exterior wall construction, not even excepting brick.

As to its durability, Mr. F. E. Kidder in his " Building Construction" states that " In Europe there are numerous examples of architectural terra-cotta which have been exposed to the weather for three or four centuries and are still in good condition, while stonework subjected to the same conditions is more or less worn and decayed."

When properly made, ornamental terra-cotta is impervious to moisture or to the disintegrating action of frost. The glazed skin produced by the vitrification of the mass causes the material successfully to resist climatic effects, even under the severe conditions common to the United States.

Severe fire- and water-tests have demonstrated that in ornamental terra-cotta the body of the block is not injured by the heat or by sudden applications of cold water. Delicate ornamental work may suffer by the sudden extremes produced by fire and water, and discoloration sometimes results to the exposed face. This is caused by the " craze " or small checks which appear on the surface of the material, and the entrance of fire or smoke into these crazes will discolor the block. This is often true even of plain surfaces. If no penetration has occurred, the surface may be cleaned with diluted muriatic acid. Very severe conditions, however, are necessary to produce any injurious effects, and even though the surface color be changed, the strength is in nowise impaired. The high degree of heat under which the material is burned almost precludes a test of equal severity while in service in building construction.

Terra-cotta will certainly require less reconstruction after severe fire- and water-tests than any building material, saving, possibly, the best qualities of fire-brick.

Mortars, Plasters, and Plaster of Paris.—*Cement and cement-mortar.*—In connection with the testing of cements for the Sanitary District of Chicago, a series of tests * was made by Thos. S. Johnston, Asst. Chief Engineer, to determine the value of cement in resisting high temperature, such as that to which concrete floors in a burning building might be subjected. Tests were made on a number of ordinary briquettes of neat cement, and mortar made of cement and sand, all briquettes having been previously immersed in water from four to twelve months. They were thoroughly dried for a week or more, in order to drive out all uncombined water or moisture.

---

* See *Engineering Record*, Dec. 19, 1896.

For the fire-test they were placed on a perforated sheet-iron plate, set over the gas-burner of a gas-stove. After roasting for about an hour a granite-iron pan was placed over the briquettes bottom side up, and the heating was continued from four and one-half to seven hours. They were then allowed to cool gradually to the temperature of the atmosphere, when they were broken in a testing-machine.

In every instance the neat briquettes showed check-marks after the heat exposure. In some cases the marks were very minute, and in others quite large. The breaks in the final test for strength often followed the lines of checking. No such checks were observed in the mortar briquettes.

The average results were as follows:

PERCENTAGE OF LOSS OF ORIGINAL STRENGTH DUE TO HEAT EXPOSURE.
Natural cement.................... neat, 89.3,   mortar, 1 to 1, 61.0.
Portland cement .... neat, 58.2;   mortar, 3 to 1, 69.6.

From these experiments Mr. Johnston deduced the following opinion: "It may be that concrete floors will, after a fire-test, hang together under heavy loads, but these experiments indicate that they are nevertheless weakened. The extent of weakening may depend upon the duration of the application of the heat. In any event, it does not appear that cement-mortar can be relied upon to satisfactorily resist high temperature. It may do so in some cases where the structure before heating was inordinately strong."

Another series of experiments to show the action of heat on cement was described by Mr. J. S. Dobie in the " Digest of Physical Tests and Laboratory Practice."

Three kinds of cement were used,—two of Canadian manufacture, and one Belgian,—two of them being comparatively slow-setting, while one cement set very rapidly. Tests were made on neat and on cement-mortar briquettes, 1 to 1, 2 to 1, and 3 to 1 mixtures. Two hundred briquettes were used in these tests, the ages varying from two months to four years.

The test-pieces were heated in an assay-furnace, the temperatures varying from 650 deg. F. to 1,775 deg. F. After heating, the loss of weight was noticed, due to the drying off of the water of crystallization. The hardened cement is a compound of hydrated crystals of aluminum, silicate, and calcium silicate; and since the crystals depend upon water for their formation, the removal of the water from the crystals destroys them, and the cement is ruined.

The neat briquettes showed cracks, as in Mr. Johnston's experiments. Briquettes which lost 19 to 20 per cent. of their weight, which is practically the amount of water required for proper crystallization, were virtually unable to resist any strain whatever, and in very few cases retained 10 per cent. of their original strength. Briquettes which were immediately placed in a temperature of 1,775 deg. F. showed a very great loss in strength for a very small loss of weight, but when briquettes were gradually heated, and allowed to rise slowly to the temperature of the furnace, the proportion between the losses in weight and strength was followed quite closely. This, Mr. Dobie claims, is accounted for by the fact that when the cold briquettes are placed under such a great heat, considerable internal stresses are induced, due to the expansion of the outside of the briquette before any of the water of crystallization is driven off.

After cooling, the briquettes became quite soft, so that they could be easily crushed in the fingers, especially those which had lost the most weight. Under the action of water, red-hot briquettes disintegrated completely.

An indication of the results obtained may be expressed as follows:

26 briquettes, heated suddenly, percentage of weight lost .... 1 to 19
26   "   "   "   , percentage of tensile strength lost, 35 to 100
21   "   "   gradually, percentage of weight lost......... 1 to 19
21   "   "   "   , percentage of tensile strength lost 14 to 98

As a result of these tests, Mr. Dobie gave the following opinion: " The great advantages possessed by cement or concrete as a fireproof material are its low heat-conducting power and its very small expansion under heat. These advantages, however, are entirely offset by the fact that it loses its strength under heat, is ruined by water applied during a fire, and will disintegrate after a fire, if not during the fire itself. These experiments tend to show, therefore, that the value of concrete as a fire-protecting material has been greatly overestimated, and that disastrous results may follow from confidence in a building protected with such a material."

These experiments on cement-mortar are given at some length on account of their bearing on a discussion of the fire-resisting qualities of concrete, which will be considered under a later heading.

Mortars and Plasters.—All grades of mortars and plasters, from common lime- and sand-mortar to the highest grades of patent and cement plasters, are used for fire-resisting purposes in various forms of light interior constructions. These have been called into existence by false notions of economy as to original cost and space occupied. Many of the hard mortars and patent plasters when applied to light metal frameworks and metal lathing have proved by experience to be more or less useful, according to the intensity and duration of the exposure; but the ultimate disintegration becomes simply a question of intensity and time, and the use of such constructions should be governed by a large quota of discrimination.

" As far as actual resistance to intense heat is concerned, common lime· and sand-mortar in small quantities, that is, when used for the joints between brickwork or as a plastering on a brick wall, has greater fire-resisting qualities than any other plastic material. It is not uncommon for the surfaces

of bricks to be melted and the mortar joints to be left standing out from the wall like a honeycomb."*

But lime-mortar has not sufficient strength to be used
alone in bodies thick enough to offer adequate fire-resistance.
If it could be used to a thickness of four inches or more, it
would be far superior in fire-resisting qualities to cement-
mortar.

Cement-mortar or plaster, when exposed in considerable
areas to the action of heat, is generally worthless after passing
through the ordeal, provided even it has answered its purpose
of restricting the fire. The partitions in the Home Life
Insurance Building, which were constructed of a light metal
framework, metal lathing, and cement-mortar, showed that
such constructions can *not* be considered as first-class fireproofing methods, although other cases may be cited, as in
the Livingstone Building, in which plaster partitions have
been reasonably efficient.

Plaster is also largely used with wire netting or metal
lathing for the protection of columns and beam-flanges, but
experience has shown (see Chapters III and IV) that unless
the plastering is well pressed through to the reverse side
of the netting, it will all soon drop off, especially under the
action of water. In cases where the netting or metal lath
hugs the metal member closely there is no room for the
plaster on the inner side, and as soon as the outer layer drops
off, the metal is left exposed. To be even partially efficient,
a space should be provided between the netting or lath and
the metal member, and this space should then be completely
filled with plaster.

Plaster of Paris.—The fire-resisting properties of calcined
gypsum, commonly called plaster of Paris, have long been
known and utilized. In parts of England where the stone is

---

* "What Constitutes a Fireproof Building Material." P. B. Wight,
in *The Brickbuilder*, September, 1896.

found in abundance it has been used for nearly three centuries, and in France plaster-concretes made of broken brick, wood chips, and plaster of Paris have been used for many generations.

Gypsum is a native hydrated sulphate of lime, the transparent grades being termed selenite, and the finest qualities alabaster. The common grades of gypsum are gently calcined until the moisture is fully driven off, after which it is ground to produce plaster of Paris.

The base for the mixture used in the Metropolitan fire-proof floor in this country is plaster of Paris. This construction has the advantage of great lightness, but the experiments of John J. Webster* (see also next heading, Concrete) show that Portland cement is superior to plaster of Paris as a matrix for fire-resisting floors. Plaster of Paris will also fail under the action of water thrown upon it during fire, as it becomes soft, and washes away to some extent.

Asbestic Plaster consists of a mixture of asbestic and lime. The asbestic is made from a serpentine rock, mined near Montreal, which contains a large proportion of asbestos. This is pulverized, producing the two different grades of "rough asbestic," which forms a tough coating similar to asbestos felt, and "finish asbestic," which consists of the rough asbestic reduced to a greater fineness in grinding. For use at the building, the asbestic is mixed with about 10 per cent. of freshly slaked lime of the best quality, the lime putty being thin enough to mix the asbestic without the addition of any more water. The plaster should then be stacked up for at least 24 hours, after which it may be tempered as desired for use.

Claims of great fire-resisting properties are made for this material, as well as resistance to the effects of water during fire; cracking and discoloration due to the percolation of water

---

* See Proc. Inst. C. E., vol. cv.

or acids are also claimed to be avoided. The plaster is tough
and elastic, and it will receive nails without chipping or cracking. The weight is said to be about half as much as ordinary
cement-mortar.

A test of asbestic was made in Washington, D. C., in the
presence of the officials from the Supervising Architect's
office. A room 4 ft. high was constructed and plastered on
the inside with asbestic. A severe fire was then maintained
for one and one-half hours, after which water from a 1½-inch
fire-hose was applied without injurious effects. The plaster
did not crack or drop, but remained intact.

All of the walls, ceilings, and columns of the new
Appraisers' Warehouse in New York City were covered with
a coat of asbestic, applied from ½ to ¾ of an inch thick
directly on the concrete or terra-cotta surfaces.

The great objection to the use of this material lies in its
slow drying. The plaster hardens only through the gradual
drying out of the mixture, and does not set like cement.
The time required for a thorough drying out is usually very
long, and constitutes a most serious objection to its employment.

'Concrete.—It may be safely said that no subject connected with the general question of fire-resisting construction
has been more discussed in the past few years than the ability
of concrete to withstand fire and water when used in floor
construction, or in the protection of columns or other structural members. The introduction of the many forms of concrete floors has provoked a wide-spread discussion, and the
believers in terra-cotta have often loudly decried the increasing use of concrete as a fire-resisting material.

The tests described in the previous section, as showing
the action of cement and cement-mortar under severe heat,
were given at some length because these tests have been
often quoted by the upholders of terra-cotta to prove that
concrete is unreliable under the action of fire. In fact,

opinions are quoted from both of the authors of these tests, in which such general conclusions are drawn. In the opinion of the writer, these conclusions are unwarranted.

It should be especially noted that the before-mentioned tests were made upon either neat-cement briquettes or cement-mortar briquettes, and that all of the tests were made for tensile strength only. Such experiments cannot form a basis upon which to found opinions regarding concrete.

There are great differences in the fire-resisting qualities of neat Portland cement, cement-mortar (or a mixture of cement and sand), rock concrete, and cinder concrete as ordinarily employed in building construction. A fair comparison can only be made by testing all of these materials under practically the same conditions. Neat Portland cement, and cement-mortar, may be properly judged from the tests of Messrs. Johnston and Dobie, but conclusions as to the various concrete mixtures must be had from other sources.

The most complete experiments ever made to determine the fire-resisting qualities of different mixtures of concrete were conducted by a commission appointed by the city of Hamburg, Germany. These tests were made some years ago, and quite an elaborate report of the investigations was issued in 1895.*

Tests were made on sixteen varieties of concrete mixtures. These included cement and sand; cement and gravel; cement, sand, and broken stone; cement and fine cinders; cement and coarse cinders; and cement, sand, and broken basalt. The tests consisted of exposing the samples to fire at a temperature of 1,080 deg. C. or 1,976 deg. F. for a period of several hours (3¾ hours in some cases), and then either cooling slowly or very suddenly by the application of cold water.

---

* For general results of report see " The Materials of Construction," by J. B. Johnson.

The report shows that while all of the sand, gravel, and stone mixtures, with one exception, either crumbled or showed greatly reduced coherence after the test, the cinder concretes, especially the coarse mixture, gave most excellent results. The latter "showed good coherence" and "did not suffer" by wetting while hot. In this respect the endurance of these concretes exceeded that of bricks laid in cement-mortar, as the report states that "some bricks cracked," and the mortar became "very tender and lost its binding power."

The highest degree of coherence in the concretes, particularly in the centre of the mass tested, was shown by a mixture of 1 part cement to 7 parts of coarse cinders. Fire was applied $3\frac{3}{4}$ hours. One sample was cooled suddenly, and one slowly, but neither suffered under the test.

Experiments have also been made by Mr. John J. Webster[*] to determine the relative fire-resisting qualities of mixtures of neat cement, and cement combined with sand, slag, fire-brick, stone, and coke breeze; also with mixtures of plaster of Paris and stone, fire-brick, and slag. Ten samples were tested of each mixture. The plaster of Paris base showed the most loss of original strength after heating and quenching; then came the cement and sand mixtures; while the best results were obtained from the 1 of sand to 4 of broken fire-brick, and 1 cement to 4 of coke breeze, mixtures.

Prof. Bauschinger of the Munich Technical School made tests of concrete pillars by heating and quenching. In his report of these experiments he stated that "of all the materials tested, Portland cement concrete stood the best, and ordinary and clinker bricks laid in Portland cement-mortar stood almost equally as well."

In the preceding chapter, the tests made by the New

---

[*] See Proc. Inst. C. E., vol. cv, 1891.

York Building Department were described at some length. The behavior of most of the concrete floors certainly belies the odium which has been, from some quarters, heaped upon this class of construction. These tests also bear out the Hamburg experiments in pointing to cinder concrete as the most desirable material; but even in the Columbian floor-test, where the concrete was made of cement, sand, and broken stone, without the added protection of a suspended ceiling, the test was very satisfactory from a general stand-point.

Speaking of stone concrete, G. L. Sutcliffe states that " Good concrete will resist fire to a very considerable extent. A thickness of 2 ins. or 3 ins. will effectually prevent iron from being damaged, in at least moderately severe conflagrations. The effect of fire on concrete is scarcely perceptible in ordinary fires, especially when the ceilings and floors are formed of that material; but in very large fires the concrete would split into irregular forms, but not until it became almost red-hot and was subjected to the action of cold water thrown upon it, and even then the result could not be compared to that of ordinary stone." *

In another of the New York tests a concrete arch of 4 ft. span, 3 ins. thick at the centre, supported a floor loaded with 150 lbs. per sq. ft. during a fire lasting 5 hours at a temperature of 2,300 deg. F., or beyond the melting-point of copper and cast-iron. Then, after cooling by a fire-hose, the floor withstood a load of 600 lbs. per sq. ft. without fracture, after which an isolated section of the arch 4 ft. long was loaded with a total weight of over 40,000 lbs., or to 4,100 lbs. per sq. ft. In commenting upon this test, the *Engineering News* gave the following editorial opinion: " It certainly seems that a concrete arch which can carry a load like this without ' laying down' after such a fire- and quenching-test

---

* See " Concrete, its Nature and Uses," by G. L. Sutcliffe, p. 187.

as described above, is entitled to rank as first-class fireproof construction.''

It has been stated that different mixtures of concrete possess varying degrees of fire-resistance, the best results being obtained through the use of *cinder concrete*. But cinder concrete itself may be made of a good, bad, or indifferent grade of cinders, and it is to be feared that the cinders ordinarily employed in such constructions are generally bad, sometimes indifferent, very seldom good.

Cinders, as bought and used by most fireproofing companies, are usually nothing more than ashes obtained from the burning of " slack,'' or the screenings obtained from large soft coal. This fuel is ordinarily used in the large power-plants, factories, or sugar-refineries from which the cinders are generally bought. Such cinders are always sold unscreened, and they are usually a very poor product—too poor for use in anything better than a cinder concrete filling over the arches proper, and even for filling a better grade of concrete is desirable.

A really good cinder can be obtained from hard coal only; but if this cannot be had, lump soft coal is far preferable to the " slack,'' or coal screenings. Cinders should always be screened through a 2-in. mesh, but this is seldom done in practice.

It is largely on account of the poor grade of cinders obtainable that the Columbian Fireproofing Co. use crushed stone as the base for their concrete. They claim that crushed stone, although possibly inferior in fire-resisting qualities to good cinders, is still superior to the uniformly poor grade of cinders employed in such constructions.

From what has been said, the author does not intend to claim that concrete can retain its *full* strength after severe heating and quenching, nor that it will not suffer some damage—possibly enough to require partial reconstruction. Hardly any material could be expected to avoid these consequences. Concrete floors, in former years, have been looked upon with suspicion as the result of the general collapse of such floors in some instances where the concrete has been used in plate or slab form, in which cases the concrete has been depended on for action as a beam, with the lower half employed in tension. Also, the bottom flanges of the supporting beams have been unprotected except by the ceiling-plaster. The lower half of the concrete slabs has been weakened by heat, which, added to the sagging of the beams, has caused failure.

But where a proper experience and knowledge have been brought to bear upon the composition and use of concrete in proper form, it should prove capable of safely resisting very high temperatures for very considerable periods of time, and this with the application of cold water in large quantities.

CHAPTER VI.

PERMANENCY—CORROSION.

Importance of Protection against Corrosion.—The tonnage of structural steel used annually in the United States has steadily increased year by year. In 1897 583,790 gross tons of structural shapes were produced by American rolling-mills. With the largely increased demand for steel buildings comes the attendant question as to the life or permanency of such construction. New problems as to the enduring qualities of steel are introduced, the solution of which can be definitely settled by time alone. Past experience gained from the study of bridges or other exposed structures of metal is of great value, as showing their action and wear under the various circumstances in which they have been used ; but the deteriorating conditions which exist in steel buildings cannot be judged and determined by the cause and effect as noticed in such structures.

The permanency of steel buildings as a class by themselves cannot be defined by any specific rules ; general locality, local conditions, and constructive materials and methods are all involved in a proper solution of each individual case.

The corrosibility of metal structures is gradually receiving increased attention from those entrusted with their design. In the past, and even largely at the present time, this matter has been often carelessly dismissed with some very general clause in the specifications calling for the covering of the metal-work with one or two coats of paint of questionable value. "Such a specification may be regarded as one which simply cares little for the maintenance of the original strength of a structure, so long as it is erected." The effects of the various building materials used, upon the iron or steel members, are seldom given serious thought. By far the largest part of the tonnage of steel used in buildings of all classes is designed without the services of an engineer, and it is to be feared that the average architect knows little about, and hence does not appreciate, the importance of adequate protection against deterioration from corrosion and kindred detrimental influences.

Relation of Corrosion and Fireproofing.—In Chapter V the fire-resisting qualities of the materials ordinarily used in fireproof construction have been pointed out. Stone, brick, concrete, terra-cotta, and plasters have been considered as regards their fire-resisting properties only, but the actions of these various protective coverings upon the steel frame, inasmuch as they affect its life or load-carrying capacity, must not be overlooked. A successful fireproof structure must be enduring,—designed to resist the slow actions of corrosion or deterioration as well as sudden fire.

Perfect fireproofing requires the complete covering of all portions of the metal framework. For this purpose many materials and methods are employed. Several distinct grades of terra-cotta are in general use, each having its good or bad points, while the various kinds of stones and bricks used may differ greatly in their effectiveness as protective coverings. Concrete is made in many different mixtures, and other patented or special products are constantly being advocated as superior articles.

In selecting materials for fireproofing purposes, their influence and action upon the life of the steel frame or other metal-work employed must be given due consideration.

Causes of Deterioration.—All metals suffer a diminution of strength, however slight or slow in action, almost from the beginning of service. Any attempt to prevent such deterioration, without a true understanding of the cause, can only by accident be effective.

The usual causes of deterioration are moisture, deleterious gases, and vibrations. The corrosive influences arising from the action of outside fumes or gases, such as are often found in manufacturing towns where a low grade of coal with sulphurous fumes is burned, the deteriorating tendencies of vibration, and the initial corrosion due to the effects of moisture on the steelwork before it is covered in by the fireproofing, are matters for the consideration of the architect or engineer entirely apart from the question of fireproofing. Moisture, in its action upon the materials employed, and the gases which arise from causes within the building, will alone be here considered as having any connection with the fireproofing of metal buildings.

These deteriorating influences may arise from the following causes :

Imperfect fireproofing.

Permeability, or chemical action, of the materials used.

Leakage or radiation from piping.

Relative Corrosibility of Cast- and Wrought-Iron and Steel.—Provided ordinary precautions are taken, experience seems to show that the corrosibility of cast- or wrought-iron or steel may be taken as practically the same.

If the material is subject to considerable exposure, and not liable to be regularly inspected, "the statement seems to be justified that cast-iron is a better material to use than wrought-iron or steel. The chief point for consideration is whether the cast- or wrought-iron or steel can be made almost permanently secure by care in maintenance. The best material may become dangerous when neglected and allowed to waste away; but in works which permit of inspection or are carefully protected, wrought-iron or steel is generally much to be preferred to cast-iron, owing to the reliability and the extent of the corrosion being visible." *

The ratio of the exposed surface to the sectional area largely determines the amount of the corrosion. In this respect, the usual compact forms employed for cast-iron columns are superior to the built-up forms of steel shapes with their many joints and pieces, and connecting rivets. But in spite of the advantages of cast-iron as regards reduced area subject to exposure, and unbroken surfaces, steel presents undoubted superiority in points of strength and homogeneity of composition, and a proper care exhibited in the design may secure still greater immunity from corrosive influences. Practical and constructive considerations also tend to make steel the more desirable material, especially in high buildings where stiffness, lateral stability, and reliability become most important elements.

To present a minimum area to corrosive influences, columns, either cast or steel, should be designed in as compact a form as possible; and, where practicable, the longest dimensions should be placed at right angles to the direction of greatest exposure. The practice of using very thin columns in the exterior walls, in the form of a plate girder set up on end, as has been done in many instances, should be avoided. Such disposition of the material presents the greatest possible surface to exposure.

These considerations also point to the advisability of using some form of open column where practicable, in order that all surfaces may be surrounded; or, if a box or closed section is used, the interior should be protected as well as the exterior. This may be accomplished by filling the interior of

---

* "Metallic Structures: Corrosion and Fouling and their Prevention." J. Newman.

the column with some protective material, as described in Chapter X.

Imperfect Fireproofing.—Imperfect fireproofing concerns the methods, rather than the materials employed. The best of materials may be used, and faulty design or construction, or carelessness in the details of setting, may render useless the care bestowed upon the selection of the protective media. Poor protection is often worse than no protection, in that it covers up its own defects and allows the slowly but surely resulting deterioration.

Among the more common forms of imperfect fireproofing may be mentioned inadequate thickness, carelessness in pointing up all joints, holes left in the fireproofing for the passage of pipes, wires, tubes, etc., and the exposure, from any cause, of a portion or portions of the steel frame. These points will be considered in later chapters, but their importance is to be emphasized as influencing the ultimate life of the structure as well as its fire-resistance.

Permeability, Porosity, and Chemical Action of the Materials Employed.—The permeability, porosity, and chemical action upon the steel frame of the materials usually employed in fireproof construction may be discussed under the following divisions: Mortars, Masonry, Terra-cotta, and Concrete.

Mortars.—The permeability of matter is that property which allows the passage of moisture. Porosity concerns the absorption of moisture. In building materials which are used to surround the steelwork, it is desirable to have the least possible permeability, in order that a minimum of moisture may penetrate to the steelwork.

"The greatest porosity in cement-mortars is found with the finer grades of sand, and the least for a mixture of two of very coarse (gravelly) sand to one of fine sand. The relative permeability cannot be assumed to vary with the porosity, since a given degree of porosity with coarse sand produces a much more permeable mortar than the same degree of porosity with fine sand."*

In general, it may be stated that mortars made with coarse sand are more permeable than those made with fine sand, while mortars made with fine sand are more porous than those made with coarse sand.

The permeability of cement-mortar decreases with the proportion of cement used, and with the age of the mortar. A good Portland cement-mortar becomes practically impermeable a few months after setting.

Both the permeability and porosity of lime-mortars are greater than for cement-mortars.

The effects of lime-mortar on iron or steel work seem to be largely dependent upon the peculiar conditions attending its use. Many cases have been recorded where metal was found badly corroded after being embedded in lime-mortar, while equally authentic reports have been made tending to show that lime-mortar is an excellent conservator of iron. While any question remains as to the certainty of its action, the use of lime-mortar should be limited to those parts of a structure which are unexposed to moisture.

The preservative qualities of cement-mortar are now very generally recognized. Innumerable instances are recorded of iron having been embedded in cement-mortar for ages, without appreciable change in condition. Iron clamps laid in cement joints in the Parthenon have been uncovered, and found to be in excellent preservation.

At the time of pulling down the old New York Herald Building, which had been built for about thirty years, Architect Geo. B. Post directed his inspector to bring him the worst corroded pieces of iron found in the building. No bad examples were found, and where the mortar had been in abso-

---

* " The Materials of Construction," by J. B. Johnson, p. 591.

lute contact with the paint, the paint itself was preserved. "All the reliable evidence obtainable goes to prove that cement-mortar has the faculty of preserving iron and steel to a great extent from corrosion, and certainly of preserving from injury the paint which it covers." *

Additional evidence as to the preservative qualities of cement will be found under a following paragraph on concrete.

The thorough covering with mortar of all portions of the steel frame, where possible, is particularly emphasized in the practice of Messrs. Holabird & Roche, Architects. In their experience with fireproofing, they have found that wherever the terra-cotta shapes are so arranged as to cover the entire surfaces of the beams, girders, columns, etc., with the mortar or cement in which the fireproofing is set, practically no oxidation takes place, and that such beams, girders, and columns are in perfect condition after twelve to fifteen years. On the other hand, girders, beams, and columns that are simply protected, without having the mortar in contact with the steel, have frequently been found seriously oxidized.

Masonry.—The use of veneer walls in skeleton construction has often resulted in the employment of masonry coverings of very doubtful protection. One of the requisites in high-building design is to secure walls of less than usual thickness, on account of the attendant reduction in weight. In many cases the metal framework is surrounded by no more than 4 ins. of brickwork. With even greater thickness, the employment of the best weather-resisting materials becomes a matter of advisability, if not of necessity.

Limestone should not be employed in contact with steelwork where the presence of moisture is probable or possible. Mr. L. L. Buck, Chief Engineer of the Niagara suspension and arched bridges, states that limestone must not be used in con-

---

* Geo. B. Post in Trans. Am. Soc. C. E., vol. XXXV, p. 472.

crete which comes in contact with iron or steel, as the corro·
sion of the metal will follow if moisture penetrates. In the
anchorage of the Niagara suspension bridge, strands of the
main cables were embedded in a concrete made with limestone,
and wherever the spalls touched the wires, the latter were badly
eaten and in some cases entirely severed. If it is necessary to
use limestone, it is better to place a layer of pure cement·
mortar or an extra thickness of asphalt around the steelwork.

In brickwork, underburnt soft bricks soon decay in damp
situations. Bricks which are dense, hard, even in texture, and
with a vitrified appearance, will resist decay. The uniformity,
density, and weathering qualities should all be considered.
Some bricks contain a large percentage of soluble salts.
Efflorescence denotes a decay, which is formed by the decom·
position of the salts in the brick or stone.

Metal-work embedded in brickwork or masonry is not
necessarily free from air and moisture, nor is it proof against
corrosion simply because it was cleaned and coated with
asphalt or paint before being covered in. For weather in·
fluences and vibration will cause the mortar in the joints to
crack and open, and each new fissure becomes an added con·
duit by which moisture, air, and other corrosive influences
can reach the metal. A constant pointing of the joints is
necessary for the lessening of deleterious action. "How·
ever, if ironwork is free from any corrosion when placed in
position, and is properly cleaned before it is coated, and is
fixed in air-tight, damp-proof, and water-tight brickwork or
masonry, it is unlikely to corrode appreciably." *

Terra-cotta.—The density of terra-cotta is the important
factor in determining its porosity and absorption. While the
absorption for different grades will vary under the same con-

---

* "Metallic Structures: Corrosion and Fouling, and their Prevention,"
by J. Newman.

ditions, the evaporation of the moisture from the material
will vary in about the inverse ratio.

Thus very hard-burned terra-cotta will absorb about 5 per
cent. of its weight when immersed in water.   Merchantable
hard-burned terra-cotta absorbs about 13 to 15 per cent. when
similarly treated.   Very porous soft terra-cotta absorbs from
30 to 40 per cent., and merchantable terra-cotta about 25 to
30 per cent.   But, while the hard stock absorbs about 13 per
cent., and the porous stock about 25 per cent. or twice as
much, the hard stock will either require about twice as much
heat, or the same amount of heat and twice as much time, to
permit the evaporation of the lesser amount of absorbed
moisture than is required to evaporate the larger quantity of
absorbed moisture in the porous stock.

This is accounted for by the fact that in hard-burned terra-
cotta the air-channels are smaller, and, as the material is of a
laminated nature, the cells in the blocks run lengthwise, and heat
cannot easily penetrate the surface.   Also, the heat, in coming in contact with the hard smooth surface, is reflected, and
the interior of the block is not affected as readily or as much
as is the case in a more porous material.   In porous terra-
cotta, reverse conditions are found.   The material is of a
granular nature, instead of laminated, the air-channels extend
in from the surface, and the surface, being neither hard nor
smooth, tends rather to absorb heat than to reflect it.

To evaporate the absorbed moisture, or to permit the air
to take it up, the air must be considerably warmer than the
temperatures of the material and the absorbed moisture, and any
conditions which would aid the heat to equalize the temperatures would hasten evaporation, while reverse conditions
would retard evaporation.   This would particularly apply to
the materials if used in a spherical shape, as this form gives
the greatest surfaces for equal volumes, and hence the great-

est exposed and evaporating surfaces, thus aiding the heat to reach all parts of the material.

But when, for instance, a column is encased with either of the materials, a new condition obtains to lessen materially the effect of the heat. As the material is applied as a casing, a great portion of the surface has been placed either against the column, or against an adjacent block of terra-cotta, and the effects of the heat and the consequent evaporation are accordingly reduced.

The elimination of all possible moisture is desirable, both from considerations of health and corrosive influences. It is therefore important that the absorption of the terra-cotta be reduced to a minimum, thus giving less evaporation to provide for. From this point of view, hard-burned terra-cotta is preferable, providing fire- and heat-resisting qualities are ignored. But aside from the absorption of the material after being placed, it is to be remembered that a considerable quantity of water is used in the setting, and although the cement-mortar will use a portion in crystallization, a surplus remains, and largely upon the inner or cooler side. The hard stock will neither absorb it nor permit evaporation. With porous stock, the moisture is soon removed.

Another element which is important concerns the neutralization of the inner and outer temperatures. As the interior of the mass is cooler than the exposed surfaces, and assuming in all cases that the air is perfectly dry, the effect of the heat from without is to neutralize the temperatures at some point beneath the surface. At this neutral point evaporation will begin. If the material is hard, this point is much nearer the surface than in porous stock, and this makes the conditions of dampness more permanent in the hard material.

Summarizing the foregoing, it is seen that a very porous material is much to be preferred as an insulation against dampness, and this, independent of the superior fire- and water-resisting qualities which this material possesses.

The employment of terra-cotta casings around the columns placed in the exterior walls, between the metal and the masonry, as now called for in the best examples of work, will undoubtedly add to the life of the columns.

Concrete.—The fire-resisting properties of cement-mortars and concrete have been discussed in the previous chapter, and it was there shown that cinder-concretes gave the best results. As regards the combination of metal-work with concrete, there is a decided difference of opinion at the present time as to whether concrete (used in its broadest sense, as indicating a mixture of cement with almost any base) acts as a preservative, or as an agency to promote corrosion. It has been often stated and much emphasized during the past few years that concrete acts as a preservative to the metal-work embedded therein, and almost any mixture which contains cement is now apt to be considered as satisfactory from the standpoint of corrosion. Instances were cited in a previous paragraph regarding mortars to show that cement-mortar has been found a most perfect conservator of iron or steel, and the conclusion is generally formed that almost any mixture of concrete must be equally as satisfactory because made with cement.

In an editorial under date of July 1, 1897, the *Engineering News* gave the following opinion : " We believe it is a fact about as well established as anything in engineering science that iron or steel which is encased in mortar of either lime or cement will be permanently preserved from corrosion. But engineering knowledge on this subject is not confined to theory. Engineers know from thousands of recorded instances that iron or steel embedded in mortar is preserved for many years without loss of section. And, knowing this, it appears to us like infinitesimal hair-splitting to question the durability of a combined concrete and metal floor on the ground that the metal embedded in the solid concrete may rust out in course of time."

The author believes that the above statement should have been limited to cement-mortar, and to concrete which does not have limestone or cinders as a base.

The effects of lime-mortar and limestone upon steelwork have been shown to be very injurious under the action of moisture ; and if Mr. Buck's observations at the Niagara suspension bridge are to be given weight, limestone should certainly not be used in concrete, unless under the assurance that each unit of limestone is surrounded entirely by cement.

As to cinder concretes, the *Engineering News* itself, in an editorial dated March 25, 1897, states that no experiments have been made to show the effects upon metals of the cinders used as a substitute for stone in concrete, adding that when boiler-ashes are moistened, the alkali, and possibly some remaining sulphur, have a corrosive effect upon the iron ; but that it is not known whether this would result from clean cinders embedded in hydraulic cement.

In the report of the Hamburg Commission, from which quotations were given in the preceding chapter, it is stated that of the many concretes tested, the one found to be nearest fireproof was a mixture of one part cement to seven parts of coarse cinders, without sand ; but that, unlike the other concretes tried, it was found to protect the iron for little more than two years, as the iron rusted, and this action was hastened by the acids contained in the cinders.

" Iron embedded in properly made and mixed water- and air-tight Portland cement concrete has not yet been shown to rust, and the preservative effects of such concrete may be considered to be established, provided the surface of the metal was clean and dry on the Portland concrete coating being applied, and free from corrosion ; and as the expansion of cement and iron by heat are nearly the same, there is no struggle between the substances to cause cracks, fissures, scaling, or disintegration."

"The iron or steel should be completely surrounded by Portland cement concrete of an impermeable character. For this purpose, as the impermeability and not the strength of the concrete is particularly required, a 3 of fine, dry, clean sand, to 1 of Portland cement, or a 2 to 1 mixture, can be adopted, a poorer concrete not being suitable." *

From the foregoing it appears that, in order to keep the iron free from rust, it should be perfectly embedded in a rich cement-mortar. This is done in the Monier system by covering the wire netting with a layer of pure cement. The balance of the arch, or column casing, or other form, could be made of cinder-concrete.

Moisture from Pipes.—All piping, whether supply, waste, or vent, should be kept entirely separated from the steelwork and outside of the fireproofing. Leakage from water-, waste-, or steam-pipes would soon cause dangerous corrosion if the moisture reached the steel. Sewer-gases are also to be guarded against as emanating from vent-pipes. Additional considerations of piping in a fireproof building are discussed in Chapter VII in connection with the installation of the mechanical features, and in Chapter X as relating to column protection.

---

* "Metallic Structures : Corrosion and Fouling, and their Prevention," by J. Newman.

CHAPTER VII.

FIRE-RESISTING DESIGN.

As was pointed out in Chapter I, a successful fireproof building must combine in its design the fire-resisting detail of all structural portions of the building, adequate equipment to cope with either exterior or interior fire, and, what is much more frequently overlooked, the proper planning or design of the general features.

A careful or scientific fire-resisting plan or arrangement of the whole is seldom given the consideration which should be bestowed upon this most important element of fireproof construction ; for without a proper general design, the attention paid to purely structural questions, or to equipment for coping with fire, may be rendered of little avail.

At a recent meeting of the National Fire Protection Association, the President, Mr. U. C. Crosby, made the following statement :

" In the poor construction of buildings lies the very foundtion of our trouble, and is to be found the direct cause of most large fires and conflagrations. If one defect alone—open communications between floors in mercantile and manufacturing buildings—could be eliminated, the loss ratio would be very greatly reduced. If you will consider the conditions connected with most large fires you will note that the rapid spread of fire was promoted or accelerated by open light- or elevator-shafts, or other vertical openings, also that badly arranged flues, poorly located heating apparatus, concealed spaces, and many other evils are frequently developed in event of fire which could have been easily prevented, many of them at little or no expense to the owner if taken into consideration by the architect in making his plans and specifications before the construction of the building. The question of proper protection against fire seems rarely to be considered by the architect. Practically all buildings, excepting a few isolated special hazard risks and new structures within the fire districts of large cities having serviceable building laws, contain a few or many vital structural defects."

"I do not refer simply to monuments of the past, but include structures now in process of erection."

Fire-resisting design will largely depend upon the following considerations:

> Materials employed.
> Subdivision of large areas.
> Interior light-shafts.
> Fire-doors and shutters.
> Stair- and elevator-shafts.
> Stairways.
> Installation of mechanical features.

These considerations are largely independent of the fire-proofing system which may later be adopted for the floors, columns, partitions, etc., and also independent, in large measure, of the consideration of the equipment to be provided for added protection.

Materials Employed.—The materials contemplated for use in the general design should be those which are known to possess fire-resisting qualities. In Chapter V such materials have been shown to be brick, terra-cotta, and certain forms of concrete.

It will not be enough to provide for the use of fire-resisting materials in the important structural features only. It must be the aim to reduce the combustible materials in the building to a minimum, whether such materials are employed for important structural members or for less important non-load-carrying portions. Floors, walls, and columns may be constructed without fault, and still the introduction of many non-fireproof partitions, counters, shelving, wainscoting and the like may so increase the volume and intensity of fire that even the best of materials will suffer damage.

No general rules can be laid down whereby to accomplish the greatest reduction of inflammable material, but each building must be a problem unto itself, in which this object will be quite as important as the structural design of the skeleton framework.

In the matter of store-counters, for instance, these fixtures, which are usually made of highly varnished and inflammable woodwork, could be constructed of a wood core, covered, except the polished top, with sheet metal, as is now applied to fireproof doors, door-frames, and window-sash. For shelving under the counters, gaspipe or cast uprights could be used, with shelves of wire netting. This detail would cost little if any more than the ordinary counter, and it would present a good appearance, while greatly reducing the fire-risk inherent in such fixtures.

In addition to reducing the quantity of combustible matter employed, the possibility of making materials non-inflammable should be considered, or the possibility of increasing the effectiveness of materials not ordinarily classed as fire-resisting. Such are " fireproof wood" and wire glass.

Fireproof Wood.—Although generally spoken of as " fireproof," the wood in question would be better described as "non-inflammable." This result is accomplished by means of a process consisting of,

First, saturation with aqueous vapor.

Second, evaporation by inducing a vacuum.

Third, charging the pores or cells with a solution consisting
of phospnate of ammonia and sulphate of ammonia.

Fourth, kiln-drying at low temperature.

This system renders the wood non-inflammable, and adequate tests have shown that when the treated wood is exposed
to flame it chars, and if the flame be continued, becomes completely carbonized. It will not take fire, nor retain the flame
when removed. Tests of strength show that the wood is
rendered somewhat brittle by the fireproofing treatment. The
effect is claimed to be permanent, and to have no injurious
action on adjacent metals, provided the wood is thoroughly
dried. The process darkens the color of the wood, and makes
it somewhat heavier, and consequently harder to work.

The following woods are regularly treated : mahogany,
cherry, butternut, walnut, red oak, sycamore, maple, birch,
pine, ash, spruce, elm, and whitewood. The cost of the process
is the same for all woods, being $50.00 per thousand feet, board
measure, exclusive of the final kiln-drying.

The use of fireproof wood is said to be contemplated by the
U. S. Naval authorities, as a result of the experience gained at
Santiago. This would apply particularly to the woodwork about
the superstructure. The U. S. Navy Department requirements
are that " when dry, and subjected to a heat of 600 degrees F.,
it will remain non-inflammable, and absolutely safe against
spread of fire from point of contact." *

In particularly important or valuable structures, or even in
the more exposed portions of ordinary fireproof buildings, non-
inflammable wood could be used to advantage, provided the
use of wood were particularly desired. The cost of the treatment would undoubtedly preclude its use in general practice.

Wire Glass is made in panes or sheets, formed upon
wire netting embedded in the glass. Heretofore a great objec-

---

* See *Engineering News*, Nov. 17, 1898.

tion to wire glass has been the cracking which resulted from changes in temperature, especially when used in large sheets, but the manufacturers of this product claim to have remedied this defect, so that its use is now becoming greatly extended.

The combination of the wire and glass forms a material which will admirably resist sudden or prolonged heat, even when accompanied by sudden cooling, as results from water thrown by a hose on the glass when highly heated. The fusion of the glass occurs at a lower temperature than does the fusion of the wire, and glass being a poor conductor of heat, each protects the other. When the glass cracks, the fragments are retained in position by the wire netting until the pieces are fused or cemented together again. Tests have shown that such glass will even melt before allowing the passage of flames. Extremely severe heat, though perhaps often occurring in a large fire, cannot be of extended area or of long duration.

The use of wire glass would seem particularly applicable to skylights, area windows, or to the windows of warehouses and other structures where fire-resistance is needed, but where appearances are not important. The wire glass now in the market is hardly suitable for use in the windows of office buildings, hotels, or apartment houses, etc., where the windows are a means of vision, as well as sources of light; but the glass companies have appreciated this, and they are now preparing to furnish a wire glass in which the wires are so fine as not to affect materially the appearance or transparency. Silvered wires could be used in good patterns, adding to the attractiveness of the material, rather than marring it.

Subdivision of Large Areas.—The subdivision of building areas depends, almost always, upon the business requirements of the occupants. Here, as in many other points connected with fire-resisting design, the interests of the owner and occupant are at variance—the result being that the owner is usually forced to concede the demands of his tenant. Wholesale and retail stores, warehouses and factory buildings, are very apt to have large undivided areas. In retail stores especially, the favorable impression on the customer is supposed to be in direct proportion to the unobstructed area, while in certain classes of warehouses and manufacturing buildings, convenience in handling goods and the arrangement of machinery are considered of more importance than the safeguard provided by efficient fire-walls.

The volume and intensity of a fire, and the rapidity with which it will gain headway, are vastly greater in very large areas than in small ones. The difficulty of surrounding and effectively fighting a fire of large area is also a much more serious matter for the fire department, and for this reason a maximum undivided area in city buildings is usually prescribed by the city building ordinances and by the fire-insurance companies.

The schedule of rating for a fireproof building, used by the Boston Board of Fire Underwriters, is as follows :

Area.—5,000 sq. ft. to 10,000, each 1,000 in excess of 5,000, $\frac{1}{4}$ of 1 cent. If over 10,000 sq. ft., each 1,000 in excess of 10,000, 4 cents (not exceeding a total of 40 cents). If building is occupied exclusively above grade floor for offices or dwellings, no charge for area. (*Note.*—If mercantile building exceeds ten stories, double area charge.)

No limitation need be prescribed for office buildings, as these are, of necessity, subdivided by fireproof partitions into relatively small office areas. This is also true of apartment houses and hotels. But in modern warehouse construction large undivided areas are often considered necessary or advisable, and these are as are commonly filled with large quantities of highly inflammable merchandise. But it is very doubtful if any great interference to business interests would result from municipal regulations which would prohibit, under any conditions, undivided floor areas in excess of 10,000 square feet. If areas larger than this were required to be divided from the ground up by solid fireproof partitions, the fire departments could hold fires in better check, and make conflagrations impossible.

It may, therefore, be stated that for fire-resisting design excessively large areas should be avoided: the more fire-partitions used, the better. The subdivision should be made either by solid brick walls, which are to be preferred, or by other effective fire-resisting partitions (see Chapter XI). All openings connecting such areas should be provided with approved fire-doors where possible, but even partially fireproof doors, if closed, are not to be underestimated in preventing the rapid spread of fire.

Interior Light-shafts.—The introduction in large store buildings of open courts or light-shafts running through all stories, and thus connecting successive floor areas, has always been considered a dangerous fire-risk. Sucn a light-shaft practically makes one room of the entire building; and more than this, as soon as the heat breaks in the skylight which usually covers the court or shaft, an immense flue is at once created. Examples of this design are to be found in nearly all large cities. Marshall Field & Co.'s retail store in Chicago, A. T. Stewart's store, now Wanamaker's, in New York, R. H. White's store in Boston, and many others could be cited as familiar examples of this construction. Fortunately this feature is not now introduced as extensively as was the custom when the above-mentioned buildings were designed, but the case of the Horne Store Building, in Pittsburg, shows that its use has not been completely done away with. In the appraisers' report on this fire, the following reference is made to light-shafts as there employed:

"In regard to light-shafts, where it is necessary to have them open with only a balustrade, as in stores, they can be protected by an asbestos covering made to roll up and come together in the centre. In office buildings and flats they can be provided with the regular standard wood shutter, or the rolling steel or the asbestos shutter. We think we have now touched a tender spot in the man who intends to build, and that is his pocketbook; he is saying he cannot afford to put on all of these protections. We are willing to admit for the sake of argument that to many these might seem, as they call them, an expensive luxury and of little use, as they are sure to be opened or out of repair when a fire occurs, and that it will take an extra man to look after them and another man to watch him. While perhaps this may be true, what would you think of a banker who leaves his vault-doors open at night and expects his money to be safe in the morning? The careful merchant or manufacturer is likened unto the virgins that had their lamps filled and burning."

The Horne Store Building is hardly a good example of the evil effects of an interior light-court, as in this case the fire entered all the floors by means of the windows at one and the same time; but even in this case the shaft undoubtedly served greatly to increase the heat in the upper stories. Had the fire entered one of the lower floors only, it is reasonably certain that the open shaft would have served to communicate the flames to all stories above, and thus have been the means of destruction to the entire building.

The appraisers recommended that light-shafts be protected by wood shutters, or by rolling steel or asbestos shutters. Fireproof shutters or hatches, closing down horizontally, have been used in wholesale stores or in factories, where appearances were not considered of vital importance, the hatches being always closed at the end of business hours. These, or rolling steel or asbestos shutters, may be of great value in preventing the passage of flame, but the author believes that reliance can be placed on such methods only if smoke or water damage on the stock or contents would not be a very serious matter. It would be almost impossible to devise a horizontal shutter which would prevent the passage of smoke or water, and these items are quite as important from an insurance standpoint as actual fire. If the fire was on an upper floor, water would leak through to the lower floors ; or if the fire occurred in a lower floor, the smoke would certainly damage the stock in the upper stories.

Rather than attempt a questionable cutting off of interior shafts at each floor, such shafts should be altogether eliminated from the design.

Fireproof Doors.—Openings between connecting buildings and openings to boiler-rooms are usually required by the municipal building laws, or by the insurance requirements, to be provided with fireproof doors. Doorways to fireproof stairways must also be fireproof. But in addition to these cases specified by city ordinances or by the insurance interests, the general use of fireproof doors is to be recommended in a fire-resisting structure.

"A fireproof door is not intended for a wood partition, neither is a wood door intended for a fireproof partition." If sufficient attention is paid to fireproof design to erect approved fire-resisting partitions, there can be no excuse for setting wooden doors and wooden door-frames in such partitions, if a proper substitute can be provided.

By fireproof doors, tin-covered doors have heretofore usually been meant. These are made of two thicknesses of tongued and grooved $\frac{7}{8}$-in. boards, laid at right angles to each other and nailed with wrought-iron nails, driven flush and clinched on the reverse side. The sides and edges are covered with sheets of tin, locked together and secured in place by narrow cleats of zinc. Nails driven through the sheets should not be allowed. The doors are usually hung in rabbeted frames made of angle-irons, with a cast- or wrought-iron sill or threshold. For openings in a fire-wall separating one building from another, two fireproof doors should be used,—one to take the brunt of the attack, and the other to hold back stray streaks of flame.

When for any reason swinging doors cannot be used, sliding doors of similar construction may be substituted. These should be large enough to project at least 3 ins. beyond the opening on each side and at the top. They are hung from the top by hangers with iron wheels running on an inclined track, so pitched that the doors will, when closed, rest firmly on the iron door-sill. At the lower corners, flaring dogs or guides are riveted to the angle-iron frame to keep the door from swinging out at the bottom. When such doors are to open during the day they should be held back by a hook provided with a fusible link. This link should come within the opening, so as to be readily affected by the transmission of heat through the passageway.

Swinging doors, including the fastenings, do not usually cost much more than one-half as much as sliding doors, and they are preferred by underwriters, as they make closer joints.

Another type which has been extensively used consists of two layers of corrugated-iron, each layer built into a substantial angle-iron frame, the two being connected by bands and angles so as to have a two-inch air-space between the layers. For severe exposure the space between the layers is often filled with asbestos or other non-conducting material. This style of door is not to be recommended.

Any extensive use of tin-covered or corrugated-iron doors has always been objected to on account of the unsightly appearance presented, and the want of a fire-door of a character suitable for general interior use has long been felt.

A metal-covered door of excellent appearance has been introduced during the past few years, and its use is largely increasing for isolated openings, and also for all doors and door-frames in buildings intended to be fireproof throughout.

The door is made of a wood core of the desired design, with asbestos paper placed in each panel recess. A fireproof paper is then applied to all of the wooden surfaces, over which is placed sheet steel or sheet copper, which is pressed in a hydraulic press to give the proper form for fitting over the panels, mouldings, etc. This metal covering for the door is made in two sheets or pieces, one for each side. When they are fitted over the door, the turned-down edges of both sheets are pressed into grooves in the edges of the wood core. The grooves are then filled with steel bands, extending all around the edges, the screws used going through the band and both sheets of metal, binding them securely in place. The jambs and casings are also fireproofed and metal-covered in the same fashion, making a door which can be relied upon to withstand a pretty severe fire for a considerable time, or long enough to allow the contents of a room to burn without having the flames pass the door.

In tests, these doors have been subjected to severe heat for about one and one-quarter hours, after which they were still intact and in shape, except that the metal covering had bulged out somewhat in the centre, but remained fastened to the edges. The woodwork inside was found to be charred to a depth of about ¼ in.

For exterior use these doors are covered with pure sheet copper, either natural finish rubbed bright, or with a green acid finish. For interior use plain steel may be used for painting or graining to imitate woodwork, or sheet steel, bronze-, brass-, or copper-electroplated, may be employed.

Although by far the best thing on the market in this line, in the author's opinion, further improvements are still to be desired in this patent.

The hinges should be *bolted* through the door and to the frame. Screws should not be permitted, not even for the latches. Heretofore, for door-frames, the metal covering did

not completely encase the frame, but simply covered the face
and architraves, turning in under the moulding about ⅛ in.
only. This defect is now being remedied, when required, by
completely encasing the entire jamb and frame.

Fireproof Shutters.—Fireproof buildings, as now constructed, are largely proof against serious effects from fire,
either internal or external, when such fire is of moderate
intensity and duration. Great improvements, however, are
still necessary to reduce the vulnerability of such buildings
from severe exposure attacks due to the burning of near-by
structures. This must be accomplished by presenting either
fewer openings or better-protected openings, in place of the
large areas of glass now found in ordinary practice. The best
protection can be obtained only by minimizing the number
and size of windows and doors which offer vulnerable points
of attack to external fire, and by thoroughly protecting such
openings by fireproof shutters.

Window shutters have generally been made of two thicknesses of tongued and grooved ⅞-in. boards, laid at right angles
to each other, and nailed with wrought-iron nails, driven flush
and clinched on the other side. The woodwork is then
covered on both sides and edges with sheets of tin, locked
together, as in ordinary practice in laying roofing-tin. When
made in two parts, the single boarding of one shutter should
extend over that of the other about one inch, so as to form a
rabbet, or close joint. They should be hung by firmly bolted
hinges, which are carried by means of hinge-eyes built into the
masonry.

Nails should never be driven through the tin, as the heat
will run in along the nails, and char the woodwork around
them, often to a depth of ¾ in. or more.

Corrugated-iron shutters have also had an extensive use.
Single-thickness shutters are made of sheets of corrugated-
iron riveted to angle-iron frames, with horizontal bands

riveted to the sheets as stiffeners. Double-thickness corrugated shutters are constructed of two plates of corrugated-iron, the corrugations running at right angles to each other. These are built into angle-iron frames, the two thicknesses being riveted to each other so as to leave an air-space between.

For very severe exposure the air-space is sometimes filled with some non-conducting material, the same as used for corrugated-iron fire-doors. Corrugated-iron or sheet-iron shutters will buckle under the action of heat, and, like corrugated-iron doors, their use is not to be recommended.

There can be no doubt as to the efficacy of iron-clad or tin-covered shutters, as they have proved very efficient protection in innumerable cases, among which the Pittsburg fire was not the least notable. Commenting upon the tin-covered shutters in the rear of the Horne Store Building, in the Pittsburg fire, the appraisers give the following opinion:

" In regard to protections on the outside openings, there is nothing better than the standard wood shutter covered with tin, as was used on the rear of the Store Building, as they are there to speak for themselves and show that they confined the flames and heat to the inside of the walls; for if they had given way they would have communicated the fire to the wood structure in the rear. But as these are unsightly on the front of buildings, rolling steel or asbestos shutters could be used, a number of kinds of which are on the market."

For general application, however, present methods of tin- or iron-covered shutters, or corrugated-iron shutters, are not all that could be desired, except for storage, warehouse, or manufacturing buildings, where the appearance is not objectionable. As pointed out by the appraisers of the Pittsburg fire, in the above extract from their report, hinged exterior shutters are decidedly unsightly, nor can the matter be remedied by placing the shutters on the interior of the building, as it would soon become difficult or impossible to operate them

on account of the furniture or stock preventing a rapid closing
in time of need.

It is hardly to be expected that the front windows of the
Horne Store Building should have been protected by tin-
covered shutters, as were used on the rear wall, but this is far
from admitting that some means of suitable protection could
not have been used. The means are to be found through the
use of rolling steel shutters. These could have been applied
to all windows at a reasonable cost, without detracting from
the architectural appearance of the front.

Rolling Steel Shutters.—These have been used for many
years, but their value as regards fire-resistance seems to have
been largely overlooked.

They are manufactured by many different companies in a
variety of forms, the exact details of construction depending
largely upon the conditions to be met. In some forms the
shutters are made of a single sheet of highly tempered steel,
with neither rivets nor hinges of any kind. Other types are
made by riveting together corrugated sheets of steel, with
additional metal shields along the sides for the protection of
the edges against wear and friction in the iron guides. For
use under intense heat a salamander rolling shutter has been
patented, consisting of wood slats encased in sheet metal, with
a shield of asbestos on the front between the wood and the
metal casing. The efficiency of this detail under severe conditions is very doubtful.

All of these forms may be arranged to coil in boxes located
over, under, or at one side of the opening, or in an external
pocket placed within a projecting cornice. The shutters may
be made self-coiling up to 20 ft. in width, without requiring
any winding apparatus. They are easily operated by means
of a rod which fits into a link at the exposed edge of the
shutter.

A reasonable amount of care in the design of the walls and

window-openings, etc., will allow rolling shutters to be used where desired, including even show-windows. When rolled up the shutters are not objectionably noticeable, and when down they form a solid barrier of considerable efficiency in themselves, behind which the firemen would not hesitate to enter and work. By keeping the shutters cool from within by means of a fire-hose, the most severe external fire should be held in check. For show-windows there is no reason why the value of the space for the exhibition of merchandise need be interfered with, while the added protection would be considerable.

The value of rolling steel shutters, however, will be largely dependent upon the cooling of one side by applications of water. Without this, it is more than probable that present types of rolling steel shutters would so warp under the action of severe heat as to become worthless as fire retardants, and if kept wet from within, it means the breaking of the windows which are ordinarily placed behind the shutters. This has led to the suggestion of employing a system of sprinklers, which, when placed in operation, would flood the spaces between the shutters and the window-glass, the water to be drained to the outside by means of scuppers placed below the shutters. In some cases double rolling shutters have been used, placed both inside and outside the window-sash. In the U. S. Appraisers' Warehouse, New York, the windows of the second story were provided with rolling steel shutters so arranged as to be operated simultaneously by electric connection.

In cases where fire-doors, fire-shutters, or rolling shutters are to be of service, they must be kept closed and in good repair. It would seem almost unnecessary to repeat such self-evident conclusions, and yet the experiences of firemen and underwriters show that in innumerable instances fire-doors or shutters which, in determining the fire risk on the structure, were supposed to be closed at all hours when not open for use, have been left open through neglect, or have been so little cared for as to become impossible to operate, and hence worthless as fire-retarders. The author has seen cases in which tin-covered shutters had become so rusted and thin through sagging and wearing against the sills or jambs as to keep the firemen busy for some time in putting out small fires which had worked in through openings in the tin coverings to the wood cores. Self-closing springs for fire-doors would constitute a commendable improvement.

Unprotected   Window-openings. — Fireproof window-shutters, either hinged or rolling, should be placed on all fireproof commercial buildings, where, at the close of business hours, all shutters could be closed without annoyance; but in buildings containing many tenants, such as office buildings, it would be almost impossible to place window-shutters and keep them closed. The tenants would undoubtedly object to their use, as there could be no systematic method of closing them without giving serious inconvenience to the occupants of the various offices. Some tenants would always be found who wished their offices open for night-work. From a theoretical standpoint, fire-shutters would seem indispensable, but their use is quite impracticable from business considerations.

In commenting on the fire in the New York Life Building, it was pointed out that the most serious error in the construction or equipment of this building lay in the failure to provide proper shutters for the court windows. This is a very simple criticism to make *after* the fire has occurred, and yet the omission of shutters was reasonably to be expected from the practical considerations enumerated above. It is probable that the owners of most office buildings would prefer to take large chances as to possible fire, rather than insist on the closing of the fire-shutters against the desire of the tenants, or have the continual annoyance of systematic endeavors to keep the shutters closed when once applied. This is well illustrated by the case of the Vanderbilt Building (see Chapter III), which had its exposed windows provided with fireproof shutters; and yet they might as well never have existed, as they were not closed at the time of the fire, and perhaps not since they were first placed in position.

And yet, even with exterior openings unprotected by fire-shutters, the external risk may still be greatly lessened by means applicable to all ordinary conditions.

The value of wire glass as a fire-resisting medium has already been pointed out, and this material, in combination with metal frames and sash, could be used to great advantage where fire-shutters would be unsightly or difficult to maintain. The frames and sash may be cast-iron, as has been used in some cases, or they may be of wood, covered with pressed sheet metal. The same company which manufactures the metal-covered doors previously mentioned, also makes window-frames and sash, wainscoting, base, chair-rails, etc., after the same principle, viz., covering a wood core with sheets of pressed sheet-metal which closely follow the mouldings employed, thus presenting a fire-resisting metal surface of attractive appearance.

This detail of window-frames and sash is to be particularly commended. A building erected in Boston, 1898, for the New England Telephone and Telegraph Company has every exterior window and door fitted with this patent, the finish of the metal being pure copper. The window-panes are all made of wire glass.

Stair- and Elevator-shafts—Considering the well-holes for stairs and elevators as usually provided in mercantile and office buildings, hotels and apartment houses, it must be admitted that, from the standpoint of fire-resistance, the ordinary location and design is wholly wrong and entirely inconsistent with other methods of protection which are provided with great thought and care. Fireproof floors, for instance, are carefully investigated before a choice is made for a well-designed structure, and the one is selected which seems to provide the greatest safety to the structure. It can hardly be claimed that the chief reason for providing a fireproof floor is to bear safely the loads upon it under a fire-test, for in ordinary cases a heavy slow-burning or mill-construction floor would answer that purpose. The idea is, however, to provide a floor which will be damaged as little as possible by fire and water, and, at the same time, which will prevent the communication of fire from one floor to another. Fireproof partitions are likewise introduced for the purpose of confining fire. And yet with these necessary precautions is combined the open well-hole for stairs or elevators—well-holes which, in time of fire, present the surest and most effective means of communicating flame from floor to floor, and from basement to attic. How often has terrible disaster and loss of life resulted in tenements and apartment houses, due to the rush of flame up the stairway or elevator-well! And experience shows that "fireproof" buildings are not exempt from the evil effects of this great defect in plan or arrangement.

It may be thought that, in many cases, sufficient combustible material will not exist to produce a fire severe enough to be transmitted by means of well-holes, but the examples already given in Chapter III show that this cannot be counted on. A most incombustible building may, sooner or later, be filled with highly combustible contents. In the Chicago Athletic Club Building the fire spread from floor to floor by means of the elevator-shaft and windows, and in the Pittsburg fire the stairways acted as a means of rapid communication from floor to floor. In the Livingston Building, where the stairway was isolated at each floor by a solid plaster partition, the fire was confined, internally at least, until the arrival of the firemen, who then prevented the possibility of further spreading.

It is, perhaps, difficult to reconcile the theoretical advantages of isolated stair and elevator well-holes with the practical disadvantages which result from such isolation; for when the requirements of adequate fire protection seriously interfere with the conventional architectural treatment of the interior of the building, the introduction of any methods of fulfilling such requirements is most difficult to obtain. Many office and mercantile buildings of modern design depend almost entirely upon the stairways and elevator enclosures for interior architectural effect; and if these features were to be relegated to an isolated enclosure, separated from the main corridors by fireproof doors, both architects and owners would consider that undue precautions were being taken, and that interior appearances were being sacrificed for a very remote possibility of danger. The matter is largely reduced to a question of appearance *versus* efficiency.

But in spite of these disadvantages or objections raised as to appearance, it is still indisputably true that stair- and elevator-shafts should be completely isolated from the floor corridors or floor areas by means of fire-doors, and that stairways should have an independent fireproof door to the sidewalk at the ground level. This may be accomplished by some such arrangement as shown in Figs. 31 and 32, which were suggested for use by Architect John M. Carrère, and published in the *Engineering Magazine* of October, 1892. These plans are decidedly the reverse of the usual practice, but in principle they are none the less a necessity for absolute fire protection.

Figs. 31 and 32 show double doors at each opening between the shafts and corridors or floor areas. If nothing more fireproof than thick hard-wood doors were used in this manner, they would still probably answer their purpose, as they would form a sufficient barrier to the flames to enable the firemen to work behind them in the shaft. A better door, however, would be the metal-covered fireproof door described before in this chapter. This is of good appearance, and effective against fire; and if two were used for each opening, they would isolate the shaft from almost any fire, however severe.

If it is claimed that, in retail stores and similar structures, the stairways and elevator-shafts cannot, or rather must not,

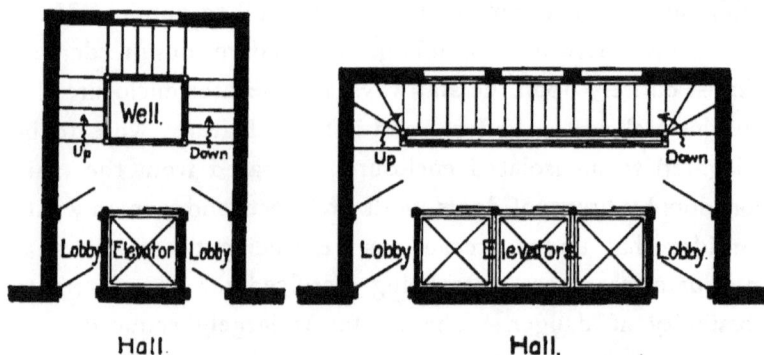

FIG. 31.   FIG. 32.
Proposed Arrangements of Isolated Stair- and Elevator-wells.

be isolated on account of commercial reasons, access of customers, etc., it is still possible to make great improvements over present methods. Elevator-doors opening directly into large areas may be made of the metal-covered pattern before described, and the balance of the enclosure, the sides and space over the doors, should be of fireproof construction. It may be urged that with such an arrangement the elevator-doors would have to be opened at each and every floor in order to see whether customers were waiting to be carried, but by using wire glass in the upper panels of the doors this objection is readily overcome.

Open grille-work in the transoms of elevator-screens should not be permitted in a thoroughly fireproof structure, unless some sort of a fireproof shutter or curtain, held in place perhaps by a fusible link, was used.

As a result of the Pittsburg fire, the appraisers recommended in their report as follows: "We would also advise that all stairs and elevator-openings be enclosed on all sides (excepting the doors) with tile protection, and with an extra rolling shutter for the doors which can be closed when not in use and at time of fire." The author believes that fire-doors are preferable to fire-shutters. The latter are apt to get out of order, and they cannot be closed in time of need as rapidly as doors, nor, when once closed, could they be readily opened for the escape of any person cut off by the premature closing of the shutter.

In elevator-shafts, strong metal grilles or gratings should always be placed under the overhead machinery and sheave-beams. This is to arrest the fall of the parts which have to be exposed to the upward rush of flame in time of fire.

To prevent the spread of flame up elevator-shafts, it has been suggested to place a system of perforated water-pipes around the sides of any shaft at the roof level, these pipes to be connected with an outside stand-pipe, with a connection for fire-engine attachment at the sidewalk level. This would form water-curtains on the four sides of the shaft, and would serve to drown out, on the sprinkler system, even a well-developed fire.

Stairs.—As the stairs furnish the only absolutely reliable means of communication between the various floors, it is necessary that the stairways be entirely fireproof, and easy of access and passage.

In an editorial on "The Need of Improved Methods for Extinguishing Fire in High Buildings," * the *Engineering News* discusses the subject of fireproof stairs so fully and well that the comments there given are quoted here at length.

"Consider first the common staircase as the means of access to the upper floors where a fire is raging. A 50-ft. section of ordinary 2½-in. cotton hose weighs from 56 to 60 lbs., and 250 ft. of 1¼-in. rope weighs 65 lbs. The carrying of either of these loads from story to story is a task which attains pretty

---

* See *Engineering News*, December 22, 1898.

serious proportions at any time, and which may, under certain conditions of inconveniently arranged stairways, or obstructions from smoke, become practically impossible, even with a fire of no very great proportions.

"That this is not a merely theoretical criticism, a review of the rôle played by the stairways in some of the recent office-building fires in New York City will indicate. In the comparatively insignificant blaze which occurred in the Vanderbilt Building on February 11th of the present year, the firemen were overcome by smoke and exhaustion several times in getting the hose from the street to the eleventh floor by way of the staircase. Neither this stairway nor the adjacent elevator-shaft were touched by the flames, and the only obstacles which the firemen had to overcome in surmounting the stairs were their crookedness and the presence of smoke. Again, in the case of the burning of the Temple Court and the Manhattan Savings Bank buildings, the slate treads of the stairways yielded to the heat, leaving the staircase with openings the full size of the tread, which made them impassable. Marble and bluestone, which are also continually used for stair-treads, have proved equally insecure when attacked by fire and water. So little confidence, indeed, is felt in such stone treads that a well-known authority on fireproof construction has recently expressed the opinion that the staircase construction in 90 per cent. of our great office buildings would prove utterly unreliable for the use of firemen in the event of a fire of serious proportions.

"These are serious considerations, especially as there is no good reason for a construction which is so insecure. The fault is due entirely to the prevalent construction, which is simply to let the stone tread into iron rabbet frames, with no support underneath them. To make the construction safe would require only that an iron plate, of openwork or gridiron pattern if desired, be placed underneath the stone tread and firmly fastened to the frame. Then if the stone broke or crumbled away there would still remain the iron plate, which would afford a safe support for the foot. Obviously, also, the staircase could be made entirely of metal, as is, indeed, sometimes done. By the exercise of a little taste the architectural aspect of the staircase would not suffer much by such a construction, and where the stairs are used so little as they are in high buildings, the greater insecurity of foothold afforded by cast-iron treads would not cut a very important figure. The metal construction would, if anything, be less expensive, and it certainly would be a great deal more reliable than the ordinary iron frame with unsupported stone treads."

Installation of Mechanical Features.—The arrangement and installation of the mechanical features of every building should be properly provided for in the original design, and not left to a haphazard installation as is only too often done. Piping of all kinds is very apt to be a serious menace to the fireproofing scheme. The pipes are often placed where they will do the most harm, and if, by chance, an attempt is made at making special provision for them, this is usually accomplished by means of open shafts, which then become admirable flues for the distribution of fire in unseen places.

If open shafts are not used, it is common to run vertical lines of piping for supply, plumbing, lighting, and heating either within the column fireproofing, or immediately adjacent to the protected columns. Provision is made for horizontal piping by means of wooden sleepers running over the floor-beams ; these are often made as much as four inches in height, to provide ample space for pipes and wires. After these features have all been placed, the spaces are filled with cinder concrete, and the finished flooring is laid. Then, even where the greatest care has been taken with the installation, repairs or alterations are made necessary by rearrangement of partitions, or different subdivisions of areas; the floors are removed, the concrete is dug out, and the work is replaced by careless or ignorant workmen who do not appreciate the damage which may be done to the fireproofing qualities of the structure. When the piping runs inside the column fireproofing, it is very common to find holes left at each floor within the column enclosure, which then becomes an active flue for the upward rush of hot air or flame.

Mr. John M. Carrère, of the firm of architects Carrère & Hastings, New York City, has suggested the following most admirable remedy for the ordinary defects of installation of mechanical features : *

" If, when a building is first planned, it is so arranged that the ceilings of the halls can be dropped, thereby obtaining a horizontal space between the floors and the ceilings of the halls on every floor, which can be lined and made entirely fireproof, all of the horizontal portion of the mechanical plant excepting the steam-heating can be placed in these ducts.

· " This process is economical both as to construction and as to installation, and every particle of the horizontal plant is then always within reach for repairs or changes, and is placed in the most favorable position for immediate connection with the rear end of all the offices, where all the plumbing fixtures, etc., should be located, instead of occupying the most valuable and well-lighted space—as is so often the case—in the front end of the offices.

" By also placing in this duct a system of air-ducts, office buildings can be economically and efficiently ventilated, either by exhaust or by indriven air, at virtually little expense. But in addition to all of these advantages, a plant thus arranged would be virtually permanent, for, no matter what changes occur in the redistribution of offices, nothing would ever have to be removed excepting the fixtures above the floor or below the

---

* See " Interior Fireproof Construction," by J. M. Carrère, *The Engineering Magazine*, Oct. 1892.

ceiling, still leaving all of the main lines and their feeders untouched.

"At the ends of these horizontal ducts, vertical chases or ducts, solidly built of fireproof material,—hollow blocks or brick,—can then be provided, running from cellar to roof, furnishing the best means of installation for the vertical plant, and connected at each floor with the horizontal ducts. For a further safeguard it becomes an easy matter, under these circumstances, to partition off the vertical chases at each floor, and the horizontal chases as often as needed, with wire lath and plaster of Paris, which can easily be removed and replaced as occasion demands with plaster boards, mineral wool, or some other equally convenient and safe material. It will be seen that by this arrangement there are absolutely no vertical chases in the centre of the building."

By using this method of horizontal ducts in the corridors, and vertical chases in the exterior walls at the ends of the corridors, or at convenient interior points, a safe, economical, and altogether commendable installation could be effected. By going a step further, and covering the faces of the vertical chases with cast-iron "slot covers," panelled, and screwed in place, access to the piping could be had at any time for purposes of changes or repair, without causing any damage to the structure itself.

For the installation of the steam-piping, vertical chases may be constructed in the exterior walls alongside of the columns. "If the steel-frame construction is used, the columns can first be fireproofed, then a chase made around this fireproofing, and a second fireproof wall built around it, doubly fireproofing the column by the intermediate space or chase.

"In this chase vertical risers can be run from top to bottom, entirely enclosed,—avoiding the objection of radiation in offices where steam is not wanted,—and these pipes can then be tapped at every floor with horizontal branches, run above the

floor and directly to the radiators, which are then placed opposite to and under the windows, as they should be, admitting of
direct indirect radiation.   By this system no steam-pipes are
placed under or in the floors, where the heat is apt to injure
the floors, and also to expand the metal work in them.   The
entire steam-heating plant is easily reached; repairs, changes,
additions, etc., are made without injury to any part of the
building ; and as the offices are determined by the windows,—
these being necessarily in the outside walls,—each and every
office can be always heated, no matter how the partitions are
changed."

See also Chapter XII, in reference to pipe spaces adjacent
to exterior columns.

CHAPTER VIII.

TERRA-COTTA FLOORS.—GIRDER PROTECTION.

TERRA-COTTA floor arches are made of either "porous," "semi-porous," or "hard-burned" terra-cotta. The manufacture of these different grades of terra-cotta and their fire-resisting qualities were described in Chapter V. The construction of the various types of terra-cotta floor arches employed in present practice will now be described, and the advantages or disadvantages of the several forms will be pointed out.

Characteristics of Porous and Hard-burned Terra-cotta. —Porous terra-cotta can be easily cut, and there are grades soft enough to allow the driving in of nails or screws for receiving the interior trim of buildings, when so desired, or for fastening slates, tiles, etc., on roofs. These soft nailing blocks are usually made solid.

The quality of porous material may be ascertained by striking the block with metal, and the result should be a dull ring. If a sound is produced which indicates a crack, the block should be condemned. The texture of the material can generally be determined by the weight of the block. While lightness is an advantage, to be abnormally light is an indication of weakness.

Hard-burned terra-cotta cannot be readily cut, but must be broken. The material is brittle, and is liable to failure under shocks. In cases where suddenly applied loads are expected, porous material should be used. Under static loads, hard terra-cotta is stronger than porous terra-cotta, in comparing

147

equal sectional areas; but this difference is largely offset by the increase in thickness of the webs in porous blocks.

In deciding on the quality of hard-burned terra-cotta, the ring should be true, when struck with metal, but the material should not be too hard, as it will not give sufficient suction to the mortar used in the joints in setting. If of a smooth surface, the suction will be poor, and the blocks should be grooved or "scored" to provide a key for the mortar.

Construction of Flat Arches.—Flat terra-cotta arches are made up of two "skews" or "skewbacks" resting against the beam webs and fitting around the lower flanges of the beams, one "key" or centre block, and "fillers," "part-fillers," or "intermediates" sufficient in number to fill the spaces between the skewback and the key. In end-construction, a filler or whole intermediate block is usually considered as 12 inches long, a part filler being less than this in length. In side-construction the widths of the fillers vary, and reference must be made to the manufacturer's catalogue for the various sizes.

To find the required number of blocks for a given space, allow about ¼-inch joint to each block and diminish the length of the key by the sum of these joints. If this makes the length of the key less than the length of the smallest key made for such type of block, take out one filler and substitute a part-filler of the most desirable size. Then add the difference between the filler and the part-filler to the length of the key as assumed. Lengths are usually taken on the top line of arch.

In scheduling arch-blocks, the dimensions are usually written—width × depth × length. Thus, an 8 × 10 × 12 filler-block would mean one in which the width in direction parallel to beams was eight inches, the depth of arch or depth of block ten inches, and the length of block, measured at right angles to the beams, 12 inches.

A safe rule for determining the spans for which terra-cotta arch-blocks may be used, is that the maximum span in feet should not exceed two-thirds the depth in inches of the arch-block employed.

Tiles laid in parallel rows should preferably break joints alternately.

Side-construction Arches.—Side-construction blocks are made of both porous and hard-burned terra-cotta. In this form of arch, the voids in the terra-cotta blocks run parallel to the supporting beams. Bevelled joints are usually employed; that is, all of the joints in each half of the arch are made parallel to the side of the key, or to side of skew. Arches are occasionally made in which the joints are radial, or meeting at a common centre if prolonged. This latter form is theoretically correct and makes a better and stronger arch, but the many different shapes of blocks required for a single arch and for varying spans increase the cost of manufacture to a point which makes this construction almost prohibitory.

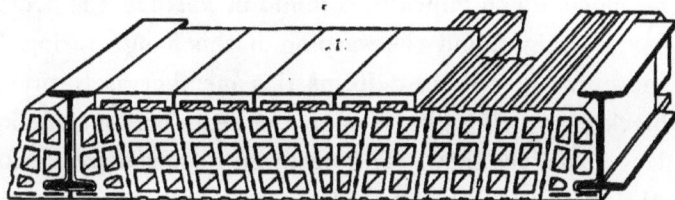

FIG. 33.—Terra-cotta Arch—Side-construction, Bevelled Joints.

Fig. 33 shows an arch of side-construction, with bevelled joints, the blocks having one vertical and two horizontal interior partitions or webs. This illustration is of a deep arch. The more shallow ones have less interior members, the size of the voids being about the same in all arches, whether deep or shallow. The better class of blocks usually has one interior horizontal web or partition in blocks 6, 7, or 8 ins. in depth, two webs in 9-, 10-, and 12-in. depth, and three or four in 15- and 18-in. arches.

The various depths of arch-blocks, permissible spans, and

average weights per square foot of arches shown in Fig. 33 are
as follows :

Depth of Arch.	Width of Span.		Weights per sq. ft. in lbs. Hard-burned.	Porous.
6 ins.	3 ft.	to 4 ft.	27	25
7 ins.	4 ft.	to 4 ft. 6 ins.	29	26
8 ins.	4 ft. 6 ins.	to 5 ft.	32	28
9 ins.	5 ft.	to 6 ft.	37	32
10 ins.	6 ft.	to 6 ft. 6 ins.	40	36
12 ins.	6 ft. 6 ins.	to 7 ft.	44	40

It will be noticed that the protection for the bottom flange
of the I-beam is made by a lip projecting from the skewback
block. This method, in side-construction arches at least, largely
displaced for a time the practice of using separate slabs of tile,
or "beam-facings," which were held in place under the beams
by bevelled lips on the skewback, as shown in Fig. 34 following.
In manufacturing the skew with the beam protection burned
on the block, much difficulty is found in keeping the projecting flange straight, and the warping of this flange during drying and burning often so deforms this member as to prevent
the block from being placed upon the beam without breaking
the flange from the skew. This occurred to such a great extent
that the majority of the manufacturers have abandoned this
style of skew, and the "beam-facing" applied separately has
now become the preferable method.

Fig. 33 shows the blocks "scored" on all sides to provide a
key for the mortar in the joints, and for the plastering.

An arch with radial joints is shown in Fig. 34. This form
is made in 8-, 9-, 10-, and 12-inch depths, weighing respectively 28, 29, 35, and 46 pounds per square foot. A 7-in. arch,
weighing 26 pounds per square foot is also made, in which the
blocks have but a single interior web.

End-construction Arches.—These are usually made of
porous material. The voids in the blocks run at right angles to
the beams, from beam-web to beam-web. The skews are made

to fit directly against the beam-web and flange, with dovetailed
lips to hold the beam-facings in place.

The blocks are commonly made rectangular in shape, with
fixed bevelled joints; never radial. The sizes vary from 6 ins.
to 15 ins. in depth. The width parallel to beams is also vari-

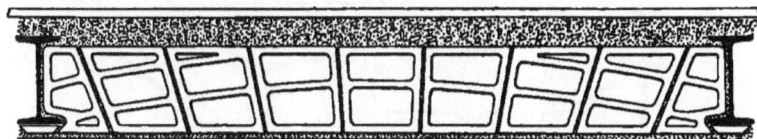

FIG. 34.—Terra-cotta Arch—Side-construction, Radial Joints.

able, but blocks of about 12 ins. width are the most common.
The lengths, at right angles to beams, are such that combina-
tions may be made to suit the variable spans between the
beams. The lengths of skews are such that the blocks of one
arch will break joints with the adjacent arches. This practice
of alternating the joints in individual arch-rings obtains in the
best classes of work. It is not the usual practice to make the
blocks break joints longitudinally or in a direction parallel to
the beams. It is not necessary that the key should be placed
at the exact centre.

For end-construction arches, care should be taken that the
arch-blocks are not too hard, for when ultimate loads are to
be carried, the blocks, if hard, will shell away at the crown
along the bearing-edges; but if the block is of a soft and tough
nature, it grinds these bearing-edges together and brings a
greater sectional area into service, thus constantly reinforcing
the strength of the arch until all of the compressive area is in
perfect bearing.

Fig. 35 shows an end-construction arch in which each arch
is continuous from beam to beam, with the blocks breaking
joints with those in the adjacent arches. This is considered
the best method of breaking joints. For this style of arch the
voids are about 3 ins. square. This arrangement of numerous

partitions gives sufficient area at all points along the line of pressure to give the required strength, and the vertical partitions occur frequently enough to give the necessary shearing-strength at the skew.

FIG. 35.—Terra-cotta Arch—End-construction.

The usual depths and spans, and the average weights per square foot for this type, set in place, are as follows:

Depth of Arch.	Maximum Span.	Weight per sq. ft.
6 ins.	4 ft. 6 ins.	29 lbs.
8 ins.	5 ft. 6 ins.	31 lbs.
9 ins.	6 ft.	32 lbs.
10 ins.	6 ft. 6 ins.	33 lbs.
12 ins.	7 ft.	39 lbs.
15 ins.	8 ft.	46 lbs.

An arch introduced by the Pioneer Fireproof Construction Company, of Chicago, is shown in Fig. 36. This form is made

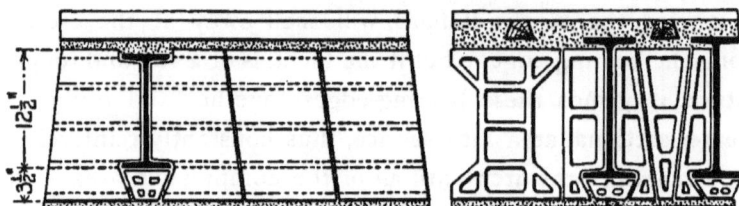

FIG. 36.—Terra-cotta Arch—End-construction, Pioneer Type.

in 15-in., 16-in., 18-in. and 20-in. depths, and is intended for heavy service. The arches are considerably deeper than ordinarily made, thereby affording additional stiffness to the floor system. This form of block, with recessed sides, possesses the advantages over the ordinary rectangular shape of permitting the tie-rods to span the bays without the cutting of the blocks, and of reducing the weight by using large voids.

Under very heavy loading, care should be taken to see that the vertical webs of the skew, resting on the beam-flanges, give the required strength.

The span lengths as given by the manufacturers are as follows :

> 15-in. arch, 8 ft. 0 ins. maximum span.
> 20-in. arch, 12 ft. 0 ins. maximum span.

Intermediate sizes are used for intermediate spans.

The weights for the different depths of arches as given by the manufacturers are as follows :

> 15-in. arch, 38 lbs. per sq. ft.
> 16-in. arch, 42 lbs. per sq. ft.
> 18-in. arch, 50 lbs. per sq. ft.
> 20-in. arch, 56 lbs. per sq. ft.

The above weights are for the terra-cotta arch-blocks only. A heavier and stronger form is shown in Fig. 37. This arch

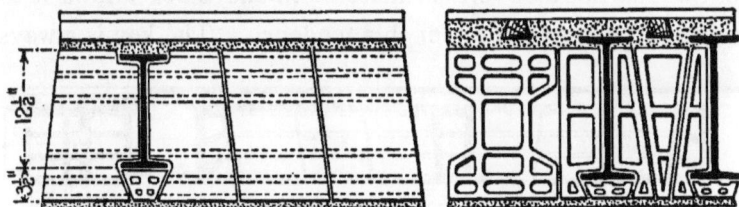

FIG. 37.—Terra-cotta Arch—End-construction, Pioneer Type.

has never been constructed, but was designed for abnormally heavy loads. The weights would average about 8 lbs. per sq. ft. over and above the weights given for the lighter form.

Combination of End- and Side-construction.—This form combines side-construction skewbacks and keys with end-construction intermediates. The combination has largely been brought about by the contractors for terra-cotta floors, on account of the large amount of mortar lost in the voids where end-construction alone was used, and because of the greater facility with which the skewbacks can be set in the side-construction.

It has been pointed out in Chapter IV that the end-construction skew is generally stronger than the side-construction skew, but it is still possible to make a side-construction skew that will be amply strong to sustain ordinary loads. This strength may be obtained by giving a sufficient sectional area to the skew to develop the required strength, and by introducing a sufficient number of partitions which follow, as nearly as possible, the lines of thrust.

The fillers in an arch loaded sufficiently to deflect or deform it are subjected to compression at the top of the arch, while in the lower portion of the arch the joints tend to open. The tendency is to wind or twist the fillers. The strength of side-construction fillers under this twisting is reduced to the shearing strength of the block, and cracking often results along the diagonal line. With end-construction fillers the large and continuous body of vertical material in the block allows it to maintain its strength under this tendency. The key is always

FIG. 38.—Terra-cotta Arch—Combination Construction.

under a compressive strain, so that it is only necessary to give a sufficient body of clay to develop the required strength, providing vertical partitions are introduced often enough to prevent the horizontal webs from buckling.

Fig. 38 shows a combination arch made by the Illinois Terra-cotta Lumber Co. This form is made in 8-, 10-, 11- and

12-in. depths, weighing respectively 27, 34, 36, and 41 pounds per square foot.

Fig. 39 illustrates the " Excelsior " combination arch, made by Henry Maurer & Son, New York. In this type, as in the two Pioneer arches previously shown, the recessed sides to

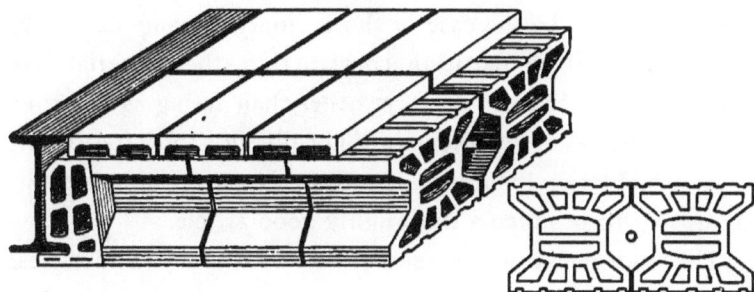

FIG. 39.—Terra-cotta Arch—Combination Construction, Excelsior Type.

the blocks provide spaces, which, while reducing the weight of the arch, leave room for the passage of tie-rods. The following spans, depths, and weights per sq. ft. are given by the manufacturers :

> 8-in. arch, 5 ft. to 6 ft. span, 27 lbs. per sq. ft.
>
> 9-in. arch, 6 ft. to 7 ft. span, 29 lbs. per sq. ft.
>
> 10-in. arch, 7 ft. to 8 ft. span, 33 lbs. per sq. ft.
>
> 12-in. arch, 8 ft. to 9 ft. span, 38 lbs. per sq. ft.

The manufacturers of the " Excelsior " arch have also patented and introduced the " Eureka" combination arch, illustrated in Fig. 40. This is especially designed for light

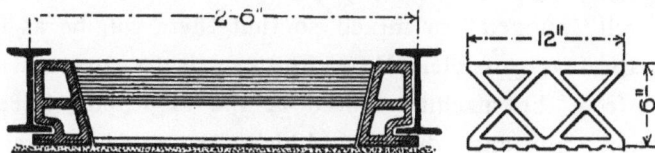

FIG. 40.—Terra-cotta Arch—Combination Construction, Eureka Type.

fireproof floors, at a low cost. Three blocks only are required for each arch, two skews, and a continuous intermediate.

Much strength is not to be expected of this type, nor is much required in residence work for which it is especially fitted.

Comparison in Methods of Manufacture.—In the manufacture of terra-cotta materials for fireproofing purposes, many of the ideas of the inventors have had to give way to practical considerations.   In the case of the ordinary rectangular block, such as is used in partition construction, the material has practically no service to perform other than acting as a barrier to heat or flame.   Such blocks are not difficult to shape, but in many arch-blocks innumerable obstacles are encountered in attempting to produce a thoroughly good article.

The side-construction skew presents many difficulties. The webs and partitions must be of equal thickness, while only the horizontal webs and partitions are of service in transmitting direct strains.   The partitions should, theoretically, radiate, or follow lines running from the top centre-point of the arch to points on the beam, which will give the result of making the voids of desirable sizes.   This means that the top and bottom webs of the skew will be horizontal, while the interior partitions will not be parallel to the webs nor to each other. The vertical webs, resting against the beams, should have rounded corners connecting with the partitions, so as to develop as much shearing-strength as possible.   These points are essential for a satisfactory side-construction skew, but they are most difficult to obtain in manufacture.

In running from the machine it is necessary to have the clay well "pugged" or mixed, so that there will be as little lamination as possible.   When the clay is too green it will come from the machine united at the angles by force or adhesion only, and when the blocks are dried cracks will appear at these points.   If the clay is well mixed, it will be as thoroughly connected at the corners as in the webs or partitions, causing no unusual difficulty in drying.   Large side-construction skews should be dried on end, thus permitting the clay to settle together while shrinking. This will relieve the tendency, which would otherwise occur, to drag the outer walls towards the centre during drying.

In running side-construction blocks from the machine, the stream of clay is moving in a line with the cells, and all of the shrinkage is made across the block, as the fibre of the clay runs lengthwise in the block. This often causes cracks on the interior horizontal or vertical partitions, which, if overlooked during burning, might destroy the strength of an entire arch-course.

In end-construction manufacture, especially where porous material is employed, the forms are simpler and the laminations in the clay do not affect the strength to any such extent as occurs in the side-construction. In end-construction blocks the buckling of the lamina is prevented by the next layers above and below, and also by the partitions. This is not the case in the side-construction method. In end-construction all the body of the arch above the neutral axis is carrying a portion of the load, while in the side-construction the horizontal members only are in direct service. This tends to produce a horizontal shearing in the lamina.

Comparison in Effectiveness.—It has been shown that, in the side-construction arch, the voids run parallel to the supporting beams. This reduces the effective arch area to the horizontal webs and partitions, as the vertical webs can serve only as sides or abutting members to the individual blocks, and as struts to prevent buckling. It has also been pointed out in Chapter IV, and in previous paragraphs, that the side-construction skewback usually constitutes the weakest place in the arch, and failure is very apt to occur at this point either by the crushing or shearing of the skew.

In the " Lee " arch, made for test at the Equitable Building, Denver, in 1890, the blocks, for the first time in terra-cotta arch construction, were made according to the end-method, and these and nearly all later tests have shown the superiority of this form. Continuous walls of material are utilized for both the vertical and horizontal webs and partitions, and all of the material performs useful work as an arch.

Raised Skewbacks.—The purpose of this form is to reduce the dead-load or weight of the arch itself, or to reduce

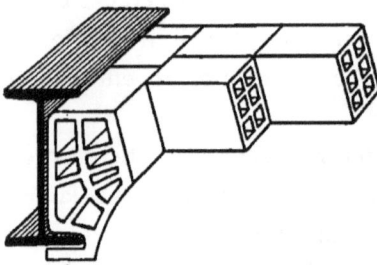

FIG. 41.—Raised Skewback—End-construction Arch.

the amount of concrete filling necessary to level up to, or cover, the beams. Figs. 41 and 42 show forms of raised skewbacks for end- and side-construction arches. These are often used in floor or roof construction, where, in consequence of deep

beams made necessary by long spans, the floor-arches can be made of a shallower depth than the beams, thereby materially reducing the load per square foot and the consequent cost.

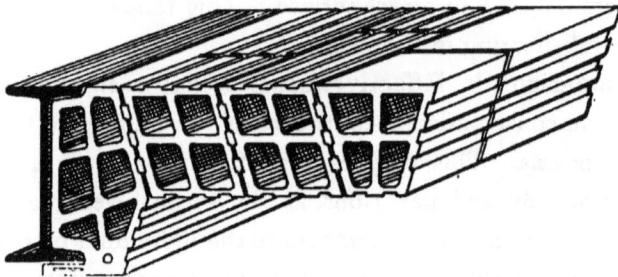

FIG. 42.—Raised Skewback—Side-construction Arch.

Terra-cotta Filling-blocks.—In cases where the terra-cotta arches do not extend up to the tops of the supporting beams, terra-cotta filler-blocks may be used instead of the usual concrete filling, thereby effecting a saving in the weight per square foot. These filler-blocks are shaped as shown in Figs. 33 and 39.

Filler-blocks as used in the Woman's Temple, Chicago, are shown in Fig. 43, but their use has never been general.

FIG. 43.—Terra-cotta Filling-blocks—Woman's Temple, Chicago.

Segmental Arches.—On account of the arched ceilings resulting from the employment of these arches, their use has generally been limited to warehouses, factories, or breweries, where considerable loads have to be carried without regard to the appearance of the ceiling. In office and store buildings a ceiling of unbroken plane is usually desired on account of the appearance, as well as for the sake of the light, which is better reflected from a uniform surface than from a broken or segmental ceiling. A level ceiling is also more effective from a fire-resisting standpoint.

Segmental arches cannot well be made on the end-construction method. They are usually made of side-construction hollow blocks, 4, 5, 6, or 8 ins. square, and about 12 ins. long. The tiles are laid so as to break joint, brick fashion. They are made of both porous and hard-burned material, and nearly all manufacturers of terra-cotta floor-shapes make one or more varieties of segmental arches. The porous material is to be preferred. Fig. 44 illustrates the usual form employed.

FIG. 44.—Segmental Terra-cotta Arch

Spans are ordinarily made from 5 to 10 ft., but in some cases 20-ft. spans have been used. They should have a rise of not less than 1 inch per foot of span, and one and one-half inches is better where practicable. The skew should **either be** very heavy, or solid, as the thrust becomes great under wide spans. The tide-rods should not be placed above the centre of the skew.

The average spans and weights per sq. ft. set, exclusive of concrete flooring or plastering are about as follows :

> 4-in. blocks, 8-ft. span, 16 lbs. per sq. ft.
> 6-in. blocks, 16-ft. span, 26 lbs. per sq. ft,
> 8-in. blocks, 20-ft. span, 28 lbs. per sq. ft.

The haunches are filled in with concrete, and this should be levelled up to a point not less than one inch above the crown. This is to prevent a direct concussion upon the blocks themselves. On wide-span segmental arches the concrete should be of good quality, as the strength of the arch at the haunches

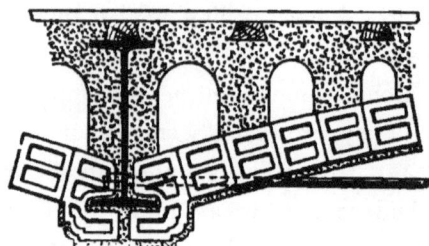

FIG. 45.—Segmental Terra-cotta Arch, with Voids in Concrete.

or end-quarter portions of the span largely depends upon the concrete, especially under eccentric loads. Voids are sometimes made in the concrete haunches, as shown in Fig 45.

FIG. 46.—Raised Compound Skew-back for Segmental Terra-cotta Arch.

FIG. 47.—Raised Skewback for Segmental Hollow Brick Arch.

Fig. 46 shows a compound raised skewback with protection-lip, for use with segmental arches 8 ins. or more in depth.

For spans not over 8 ft. a segmental arch may be built of hollow brick with raised skews, as shown in Fig. 47. This makes a very strong floor for this span.

Segmental Arches with Flat Ceilings.—The curved soffit resulting from the use of segmental arches may be concealed by the use of a suspended ceiling as shown in Fig. 48.

FIG. 48.—Segmental Terra-cotta Arch with Flat Ceiling.

This will be found both strong and economical under long spans, where heavy loads are specified; but where the spans are moderate, so that an ordinary flat terra-cotta arch may be used, the latter is be preferred.

Tie-rods.—The use of tie-rods is necessary in all these forms of flat and segmental tile arches, to take up the lateral thrust due to unequal loading on the different bays. If the arches in all bays were loaded equally at the same time, tie-rods would be necessary in the outside bays only, but such equal loading cannot be relied upon even under the most ordinary loads. Tie-rods are usually made $\frac{3}{4}$ in. in diameter, with thread and nut at each end. Rods $\frac{7}{8}$ in. in diameter are sometimes employed where extra heavy loading is provided for. The spacing of tie-rods varies from 4 ft. 6 ins. to 7 ft. centres.

If tie-rods for segmental arches are properly placed—that is, within the lower third of the beam, or preferably at the centre of the skew—their necessary exposure constitutes a serious objection to the use of this type of arch unless a suspended ceiling is used. Where a level ceiling is not employed, the methods of protecting the tie-rods against attack by fire are unsightly and unsatisfactory. Special-shaped tiles are sometimes used, giving a panelled effect to the arches, or a wrapping of metal lath and plaster is made to suffice. These details are of poor appearance, and, what is of more consequence, of small fire-resisting value.

Ceiling Finish.—The under surface of the terra-cotta flooring-blocks is usually finished by applying two coats of plaster and one coat of skimming. Many forms of terra-cotta floor-blocks are grooved or " scored " before being burned, in order to afford a rough surface for the plaster. This is indicated by Figs. 33 and 42.

If irregularities exist in the trueness of the ceiling, they may be built down to a level surface when the brown or second coat of plaster is applied. False-beam effects may be secured by the use of metal furring, as described in Chapter XIII.

Floor Finish.—Where the terra-cotta arches or filling-blocks do not extend up to the tops of the supporting beams, this space must be levelled up with concrete. The nailing-strips or "screeds" are then placed to receive the flooring, the intervening spaces being also filled with cinder concrete.

Nailing-strips are usually made of a dovetailed shape, about $2\frac{1}{2}$ ins. wide at the top, $3\frac{1}{2}$ ins. wide at the bottom, and 2 ins. thick. These are run over and at right angles to the beams, being held in place by some form of clip which secures them to the beam-flanges. The screeds are sometimes made rectangular in shape, about 2 ins. square, and fastened to the bottom is a $\frac{7}{8}$-in. board which projects about 1 in. on each side. These boards are fastened to the beams with clips, similar to the previously described shape, but the working of the screed, permitted by the shrinkage, is overcome to a greater extent than where dovetailed strips are used.

After all piping or wiring which is intended to go below the flooring is in place, the spaces between the screeds are filled with cinder concrete. This should never be omitted, as air-spaces under the flooring will largely contribute to the spread, intensity, and resultant damage of fire. This was well illustrated in the Home Life Building fire, described in Chapter III. In this case even the concrete filling between the beams was dispensed with, resulting in the failure of several beams and the complete destruction of nearly all wood flooring.

If a double flooring is used, a $\frac{7}{8}$-in. under-flooring is first laid on the screeds, upon which is placed the finished floor. If only one thickness is used, $1\frac{1}{4}$-in. matched stock is most common. Another method is to lay a 2-in. plank floor directly on the beams, secured by means of clips, over which is generally placed a finished flooring. The height of the finished flooring above the beams is made as small as possible, but it is seldom less than 3 ins.

Wood floors are gradually being eliminated in many parts of fireproof buildings. Concrete floors with a cement finish, granolithic, mosaic, and tile, are being extensively employed, but a still more common use of these floors would greatly help the elimination of combustible material.

The concrete used for levelling up between the beams and for filling in between the nailing-strips is often very defective in quality. It is generally termed " filling," and is common to terra-cotta floors and to most forms of concrete or composition floors. A light incombustible mixture of almost any variety is supposed to answer the purpose of filling in the spaces between the sustaining arch or plate and the finished floor, thus preventing a free circulation of hot air or flame. Its ability to contribute materially either to the strength of the arch or to the fireproofing qualities of the arch is not usually considered a requisite. The concrete is rather considered of secondary importance. This practice is decidedly wrong, and poor, weak, or mud mixtures should not be tolerated. Much depends upon the concrete filling to protect the terra-cotta arches against sudden blows, and to distribute properly the floor-loads, as well as the loads resulting from column-casings and partitions. The concrete is also valuable as a means of added stiffness in the floor system. The lack of lateral strength often developed by poorly constructed terra-cotta floor-arches was commented upon in the report of the engineers appointed to examine the Horne buildings in Pittsburg.

Weather and Stain Protection. — Terra-cotta floors should always be protected against rain or snow, if apt to be followed by freezing and thawing, as the mortar-joints will be injured. This would probably result in a later deflection of the arches. The blocks themselves are also weakened by the action of frost, and if long continued, are liable to crack and allow the falling of the arch.

If plastered ceilings are to be used, the terra-cotta work should be protected against the smoke or soot coming from hoisting-engines at the building. Stains are also quite liable to occur from the effects of iron in the clay, or from the cinders in the concrete over the arches, if the floor is allowed to become wet. This is especially true if the rooms below the arches are warm. To prevent these stains, several hydraulic paints have been used, some of which possess considerable merit, both as preventing stain and dampness.

Method of Setting. — For the erection of terra-cotta arches wooden centers are used. Iron clips, about $2\frac{1}{2}$ ins. by $\frac{1}{2}$ in. in size, are first hooked around the upper flanges of the I-beams, and from these clips are suspended $\frac{3}{4}$-in. diameter hooks or rods, with a thread and nut at the top for adjustment, and a hook at the bottom to support the 3-in. by 4-in.

FIG. 49.—Detail of Centers for Flat Terra-cotta Arches.

wood stringers which run at right angles to the beams. On the stringers are placed 2-in. planks, dressed on one side to a uniform thickness, laid close together, in a direction parallel to the beams. These planks receive the terra-cotta blocks. See Fig. 49.

In adjusting the centers, a sufficient camber should be given to make up whatever spring there may be to the stringers during the time of setting. This camber is usually made by laying upon the stringers between the beams wood strips sawed with a rise of about $\frac{1}{8}$ of an inch per foot.

In setting the tile it is very common to build the arches in string courses on the beams, first setting all the skews, then all the intermediates, and finally all the keys. This is bad practice, as it loads the center, both planks and stringers, to excess, causing too great a deflection. In the end-construction the arches should be built one by one, each being complete before the next is started. In side-construction, where joints are broken longitudinally, the arches should be keyed up or completed at the first point where the intermediates meet the lines of the key, thus completing the successive arches as rapidly as possible.

As the protection of the lower flanges of the steel beams is of vital importance in case of fire, great care is necessary in the placing of the skews and the beam-facings. Among builders generally, this is passed by without due attention. It is customary in setting the skew with the beam-protection worked on the block, to spread the mortar on the top of the lower flange, and then push the skew in place. In setting the skew in this manner no care is taken to see that the bottom of the beam is given any room in which to expand under excessive heat, as in most cases the protection on the skew will be in contact with the beam. When expansion of the beam does occur, the beam protection will break away and expose the steelwork.

To avoid this, the distance between the beam-protection on the skew and the seat of the skew where resting on the lower flange should be considerably larger than the actual measurement of the beam-flange itself plus the space required for a proper mortar bedding. The mortar should be spread as thinly as is practicable to give a perfect bedding for the skew. If set in this manner, the protection-lip on the skew will be at some distance from the beam, and when expansion occurs the protecting flange will not break away.

In setting the skew with the beam-facing applied separately the same care is necessary to prevent the facing from coming in contact with the beam. The beam-facings should be placed on the centering without mortar. They should be of sufficient width to come in contact with the skews, and what mortar is used in setting the skews should be high up on the bevelled lips, so that when the skews are forced in place the mortar is largely at the top edges of the beam-facings, forcing them rather away from the beam than towards it. These separate beam-facing pieces are usually made 12 ins. long, and in a variety of widths to suit the different sizes of beams used.

In securing the centers care should be taken to see that sufficient room is left between the top of the center and the bottom of the beam to permit the placing of the beam-facings as before described, and when the beam-facings are in position the center should not be raised up by tightening the bolts, as is often done. This would bring the facings into contact with the beams, and is especially to be avoided.

Intermediates and keys should be sufficiently bedded to give an even bearing, one block to another. All joints should be filled with mortar, especially at the top. The blocks must be in close contact, well shoved in place. If a space occurs at one side of a key, a solid slab of tile should be inserted, well covered with mortar; or, if the opening is too small for this, a slab of slate should be used.

Mortar.—In warm weather, when the blocks can be well wet down, clear cement mortar should be used. In cold weather, when wetting the blocks would endanger them through possible action by frost, cement-mortar should be used with the addition of enough lime-putty to prevent the tile from absorbing all of the water from the mortar.

In dry weather the centers should remain at least 48 hours. In wet weather they should remain considerably longer, depending upon the exposure to moisture.

Centering Segmental Arches.—Under the beams are hung stringers made of plank wide enough to project on each side beyond the beam-flanges, thus making a shelf on which may rest the curved centering. These are usually made of curved pieces cut out of 2-in. plank, placed on edge at intervals to receive 2-in. by $\frac{7}{8}$-in. boards, forming the segmental surface.

Where the spans are variable, but with the same rise per foot, it is best to make the centers for the widest spans. These can then be cut away at the sides and be made to fit the shorter spans.

Defects in Terra-cotta Arches.—A great disadvantage in the use of the hollow-tile system of floor-arch construction lies in the difficulty of adapting this system, or indeed almost any system requiring blocks or slabs, to the filling of irregular-shaped spaces. Terra-cotta floor-arches can be used to the greatest perfection where the arrangement of the beams and girders is rectangular; but in many cases such rectangular arrangement is impossible, due to the outline of the building site. The conditions of floor-framing often cause other irreg-

ularities, such as radiating girders, and other distortions around
elevator-wells, light-shafts, etc.   Under such conditions the
irregular panels become largely a matter of patchwork, without systematic arrangement of the blocks, due to the incapac-
ity of the rectangular blocks to adapt themselves to other than
rectangular forms.   The result is a filling of tiles and parts of
tiles wedged together in the most convenient manner, with a
plentiful supply of mortar to fill the interstices.

Details requiring Careful Inspection —In setting the
tile-blocks in these forms of floor-arch construction, great care
must be exercised to restrict the use of broken or imperfect
tile ; to prevent carelessness in opposing rib to rib in the same
arch-ring ; to secure properly mortared joints ; to protect
properly all exposed portions of steel framework, and, in general, to obtain uniformly good workmanship in all details of
setting.

The defects are often hard to detect, as the blocks are laid
on wooden centering, and while the top may appear to fulfil
all conditions of good workmanship, the bottom may appear
very different when the centering is removed.   The architect
is very apt to pass over such defective work, owing to delay
if replacement is demanded, and the excuse of "common
practice" justifies the results.

The great carelessness which may obtain in the setting of
tile arches was well pointed out in an article on " Hollow-tile
Fireproofing in the Park Row Syndicate Building," * and the
defects in setting, above enumerated, were strikingly illustrated
by photographic views taken throughout the building.   Such
defects are due to injured material and poor workmanship,
rather than to the nature of the arch material itself ; and as all
of these faults can be corrected by a more careful supervision

---

* See *Engineering News*, April 14, 1898.

of the workmen, and more careful handling and closer inspection of the material, such inspection and care become of great importance.

Selection of Arch Forms.—*Material.*—From the records of careful tests for strength made on terra-cotta arches, and described in Chapter IV, it is evident that for strength alone under static loads, hard-burned terra-cotta is stronger than the porous variety. Impact, however, must also be considered, and adequate tests show that porous terra-cotta is much superior to hard tile in resistance to shocks. But if a choice of material seems difficult under these conflicting properties, a consideration of the action of these two varieties of terra-cotta under fire- and water-tests, as described in Chapters III and IV, must show that porous tile is far more reliable and satisfactory under the trying conditions of load, shock, fire, and water. The conclusion is therefore warranted that—

1. Porous terra-cotta is much to be preferred to the hard-burned material.

Form of Arch.—Flat, unbroken ceilings are always to be preferred to any type of terra-cotta arch which may require a panelled or vaulted effect, due to the projection of the girders or beams below the main ceiling-line. A perfectly level ceiling-surface reflects more light and provides a better lighting for the areas farthest removed from the windows, and, what is even more important, an unbroken ceiling deflects heat. In a panelled ceiling the exposed area is increased where projections occur, and these are located at the very places where the most protection is needed, viz., at the beams or girders. The panelling forms pockets for the retention of heat and flame, as described under the next heading, Girder-protection.

The evil effects resulting from panelled ceilings were amply illustrated in the case of the Horne buildings, in both of which cases (see Figs. 16 and 17) the beams and the girders projected down below the soffits of the terra-cotta arches.   For proper fire-resistance, it is therefore necessary that—

2. Flat terra-cotta arches should be used so as to provide a level unbroken ceiling-surface.

End: versus Side-construction.—As to whether the arch should be of side-construction, combination, or end-construction, the considerations before mentioned in this chapter and in Chapter III point conclusively to the superiority of the end-construction method.   End-construction hollow-tile arches can be laid, under the requirements generally found in buildings, to sustain ultimate loads of 2,000 lbs. per sq. ft.; while even under the most favorable conditions the side-construction type seldom develops one-half this strength.   This is largely due to the usual form of the side-construction skew-back, which constitutes the weakest portion of the arch.   If side-construction arches are used, it is important to specify a sloping reinforcing rib in the skew-back, which shall start directly above the flange of the skew-back beam, and at a point about midway between the beam-web and the edge of the flange.

As the result of his many tests and experiments on terra-cotta arches, Mr. Geo. Hill recommends that for loads under 150 lbs. per sq. ft. total, either end- or side-construction arches may be used; but for loads exceeding 150 lbs. per sq. ft. total, end-construction arches should always be used, with the best quality of mortar.

The third conclusion may then be made that—

3. End-construction arches are stronger and more reliable than side-construction or combination arches.

Form of Blocks.—The greatest strength and heat resistance will be obtained with the least material when the blocks are made of the simplest rectangular form.   The comparison of the manufacture of side- and end-construction blocks in a previous paragraph shows the advantages of strength possessed by the simplest forms of end-construction blocks of porous material. To withstand the action of fire, however, heavy tile are much more reliable than thin tile. The tendency in some localities has been to make the material too light. Thick walls and partitions in the blocks add to the strength and also add materially to the fire-resisting qualities. If the interior angles are well rounded, breaking under fire- and water-tests will not easily occur. It may then be stated that—

4. The simplest form of rectangular end-construction blocks should be employed, with thick webs and well-rounded interior corners.

Depth of Blocks.—In arches of the same depth the strength varies directly as the span. In arches of the same span the strength varies as the square of the depth. A deep block, therefore, makes a much stronger floor than a shallower block for the same span, and, what is equally important, a lighter and cheaper floor. The floor is lighter because the additional depth to the terra-cotta block will weigh less than the concrete levelling which would be necessary over a shallower arch, and a terra-cotta arch made the full depth of the beam is cheaper than a smaller arch levelled up with concrete.

But, aside from the questions of cost and strength, fireproofing considerations make it desirable to employ a floor-arch of a depth equal to the beam which serves as its support. The fire in the Home Life Insurance Building showed the evil results attending the practice of permitting continuous voids between the tops of the floor-arches and the under side of the flooring.

Where shallow floor-arches are used the temptation must always be to save the cost of the concrete levelling, in all but the very best examples of construction. Where the arches are made the full depth of the beams, and cinder concrete filling is laid between the screeds, voids become impossible. This cinder concrete filling should never be omitted, as it greatly increases the strength of the terra-cotta arches, even though it be of poor quality, and it serves to fill the spaces between the wood flooring and the arches.

It is also necessary that the terra-cotta skew-backs shall be placed low enough down on the beams to permit the use of beam-facings which provide an air-space below the beam-flanges. In average work this is seldom obtained, and the workmanship in setting skews and beam-facings is often very poor, as was pointed out under Method of Setting.

From considerations of cost, strength and fire resistance, it follows that—

5. The arch-blocks should be of the full depth of the beams.

6. The arch should be placed low enough on the skew-back beam to allow the use of hollow or recessed beam-facings.

7. A concrete filling of good quality should be employed to strengthen the arch and to fill all voids between the arch and flooring.

Girder Protection, Importance of.—The many failures which have resulted in time of fire from the great exposure and consequent failure of girders with insufficient protection, show the vital necessity of protecting these members as carefully as ingenuity can devise. Girders, where they project below the ceiling-line, as is commonly the case, are much more exposed to the injurious effects of fire and water than the floor-beams. Intense heat is brought to bear on the corners or exterior angles of the girder-protections, and the streams from the fire-hose tend to tear off the fireproofing of the girders, due to the resistance offered in the corners. The interior angles also create dead-air spaces at these points, which cause the flame to split or separate, thus allowing the gathering of superheated air in the interior angles. This often vitiates the cement, and if the blocks are not joined mechanically they will fall from position.

Girders usually carry several or many floor-beams, or great concentrated loads, and the importance of properly protecting them must be in the direct proportion to the load-carrying functions. Questions of cost, appearance, or head-room should not be considered to the detriment of safety.

Terra-cotta Girder Protections.—Girders carrying extensive loads should preferably be provided with not less than 4 ins. of terra-cotta protection. This thickness is often avoided on account of the cost, or headroom, but the adequate protection of important members should not be sacrificed to such considerations. The bottom protection slab or block should

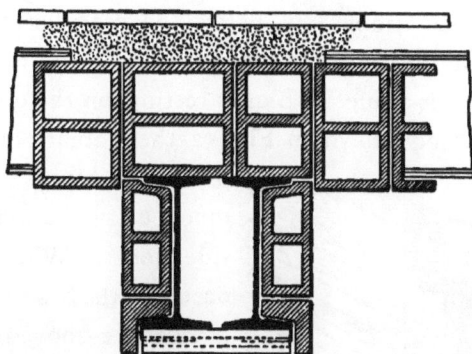

FIG. 50.—Double I-beam Terra-cotta Girder Protection.

be free from the girder, with at least a one-inch air-space, as previously noted under Method of Setting. All of the blocks used should contain as large air-voids as possible, and should be secured both top and bottom. Where metal clips are necessary they should be protected from fire by the blocks which they hold.

In Fig. 50, metal clips, one on each side, are turned into the bottom block or soffit-tile, the other ends being bent around the lower flanges of the beams.

A better detail is shown in Fig. 51. In this case ordinary .

rabbeted ceiling-tile are used for the soffit-blocks, the rabbets

FIG. 51.—Double I-beam Terra-cotta Girder Protection.

running at right angles to the girder. Passing under the blocks in each rabbet is placed a 2-in. by $\frac{3}{16}$-in. metal hanging clip, turned up at the ends over each flange. The side blocks may be made of either one or two pieces, depending upon the depth of the girder. They receive a bearing upon both the beam-flange and the soffit-tile. At the tops they are held in position by the floor-arch, which drops 1 in. below the cross-beams that rest on the top of the girder. The portion of the arch directly over the girder is made by blocks resting on the girder-flange.

In the method shown in Fig. 52 the stability of the flange

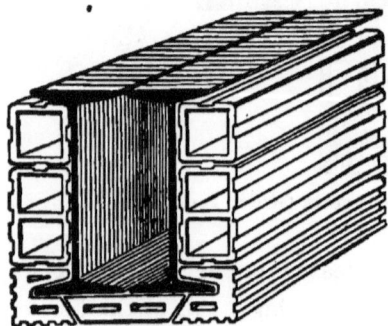

FIG. 52.—Double I-beam Terra-cotta Girder Protection.

protection-blocks depends too much upon the weight of the side blocks. When expansion occurs in the beams, the flange blocks are moved out, thus allowing the soffit-tile to drop. This is also often caused by constant vibration. The detail is also wrong in that the soffit-tile is placed directly against the lower flanges of the beams, without an intervening air-space.

For single beams used as girders, Fig. 53 and Fig. 54 represent acceptable methods. The flange-blocks are held in position by the weight of the side blocks only, but it is difficult to obtain a more satisfactory detail.

Girders which receive the floor-beams framed flush on top,

generally project only a few inches below the ceiling line. These are relatively much easier to protect. A good plan is

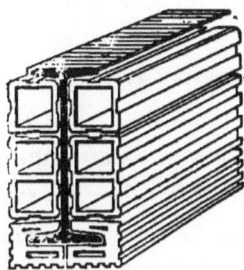

FIG. 53 —Single I-beam Terra-cotta Girder Protection.

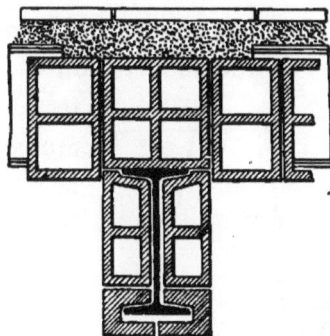

FIG. 54.—Single I-beam Terra-cotta Girder Protection.

to use skew-back blocks along both sides, with a beam facing beneath, as illustrated in Fig. 55. The same detail may be used where double beams are employed under like circum-

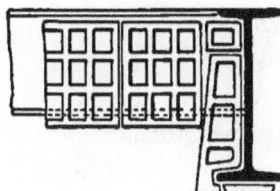

FIG. 55.—Single I-beam Terra-cotta Girder Protection.

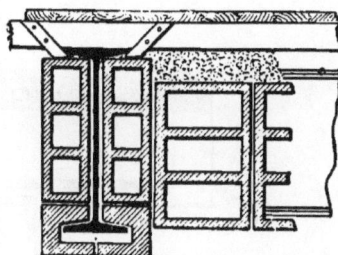

FIG. 56.—Single I-beam Terra-cotta Girder Protection.

stances, except that the skew must have a deeper rabbet to receive the wider and thicker beam-facing. Another method is shown in Fig. 56.

Protection of Plate- and Box-girders.—Practically the same principles used in beam-girders are applied to either plate- or box-girders, and the same requirements of adequate thickness in the terra-cotta blocks, mechanical fastening to insure stability, and air-spaces around the flanges, must obtain with even greater care, due to the increased loads usually borne by such members.

Fig. 57 shows the manner of protecting a plate-girder. Ceiling-tile, rabbeted on the ends, are used for the soffit pro-

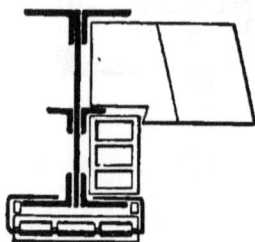

FIG. 57.—Terra-cotta Protection for Plate-girders.

tection. Metal clips pass under and around the girder angles, at each joint or rabbet in the soffit-blocks. The side blocks bear on the lower flange-angles. Small pieces of tile may be inserted between the side blocks and the soffit-tiles, or the side blocks may be rabbeted so as to extend down to the top line of the soffit-blocks. The side blocks project up beyond the ceiling line, being held in position by a rabbet in the skew-back of the floor-arch.

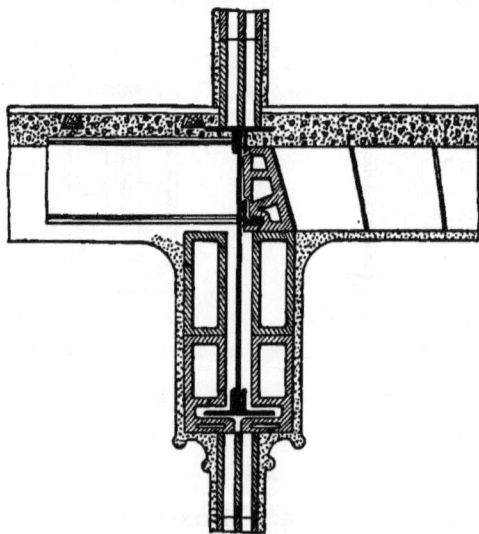

FIG. 58.—Terra-cotta Plate-girder Protection for U. S. Public Building, Portland, Oregon.

Fig. 58 shows the method of protecting the plate-girders in

the U. S. Public Building at Portland, Oregon. The same general details are employed for plate-girders with flange-plates, or for box-girders.

. The carrying of any loads by the side blocks of girder-protection should always be carefully avoided. The method shown in Fig. 57 is sometimes employed without the use of shelf-angles on the girder to support the floor-arch, in which case the skews are placed directly upon the side protection-blocks. This should never be permitted.

CHAPTER IX.

CONCRETE AND COMPOSITION FLOORS.

THE constantly increasing heights of city buildings, and the attendant extended application of steel skeleton construction, have greatly encouraged the design and development of fire-resisting floor systems, all of which aim to surpass the hitherto recognized standard of terra-cotta construction. The ends sought after have been increased load-carrying capacity and fire-resisting qualities, combined with decreased weight and cost. Up to the present time some fifty different systems of fireproof floor constructions have been placed on the market, most of them patented and controlled by construction companies.

Such floor constructions vary widely in principle; many of them are founded on sound engineering practice, while many others are open to severe criticism and should be used with extreme caution. The various systems may be combined in general classes as follows:

I. Concrete arches, with and without steel reinforcing members.

II. Concrete or composition slabs, with or without steel tension members.

III. Terra-cotta blocks with tension cables, and

IV. Dome constructions, with and without steel reinforcing members.

Class I includes short-span concrete arches, without reinforcing members of any kind, and such floors as the Roebling and Expanded Metal Co.'s arched floors, Melan, and the Clinton Wire Cloth systems with arched metal members.

Class II embraces concrete or composition beams or slabs, as in the Thomson and Mackolite floors, and slabs with steel tension members, as in the Columbian, Bailey, Rapp, M'Cabe, and Metropolitan floors.

Class III includes the Lee hollow-tile and cable-rod floor.

Class IV of dome construction is represented by the Guastavino system, without steel reinforcing members, and the Poulson, now called Hecla, floor, depending upon arched flats of metal.

Many of these floors have had a very extended use, while others have been limited in their application to comparatively few examples. The more prominent forms as recognized by architects and engineers comprise the following:

Concrete Arches without Reinforcing Members.—Concrete arches built without the aid of any metal reinforcing members are limited to short spans, and as such are not applicable to ordinary building conditions; while, even with practicable spans, the necessary depth at the haunches to obtain the requisite rise becomes an obstacle to their successful employment.

Concrete construction for entire buildings has been much more common in the Western States than in the Eastern, considerable work of this character having been done in San Francisco and the vicinity, a notable example of which is the Leland Stanford, Jr., Museum at Palo Alto, California. The hotels Ponce de Leon and Alcazar, St. Augustine, Florida, are other well-known buildings constructed of concrete throughout. But in these examples the concrete has generally been reinforced by twisted iron rods placed either vertically or horizontally within the walls or floors.

Concrete arches have been used in warehouses as floor supports, with a span of 7 ft. 4 ins., without the aid of any metal reinforcing members. These arches were 5 ins. thick at the crown and 2 ft. deep at the haunches.

Melan Arch.—The strength of this system depends mainly upon the use of steel ribs, which are usually made of tees or light beams, bent to the shape of the arch, and sprung from the lower flanges of the supporting I-beams, see Fig. 59.

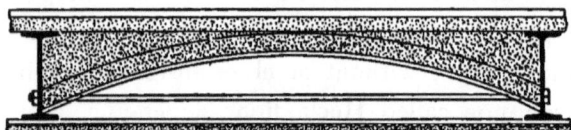

FIG. 59.—Melan Arch.

Thin steel wedges are driven between the ends of the curved ribs and the webs of the floor-beams. A curved centering of wood is applied close to the under side of the ribs, and then, starting at one skew-back, coarse concrete is filled in and rammed toward the haunches for a depth equal to about two-thirds of the depth of the curved beam. The remaining third is filled with a finer mixture of cement and sand. Each side is built and rammed separately, and the key is then filled in and rammed vertically. The haunches are next filled with a coarse mixture, either level with the tops of the beams across the entire arch, or around the beams only.

This system has been used on spans of from 12 to 16 ft. with the curved ribs spaced from 3 to 5 ft. centres, according to the strength required. The rise of the arch is from $\frac{1}{10}$ to $\frac{1}{12}$ of the span. Tie-rods are used to take up the arch-thrust.

Considerable criticism may be made against this construction, owing to the use of concrete as a beam and as an arch at one and the same time. Acting as a beam from rib to rib, the concrete must resist tensile strains, and, if the principle of allowing the use of no concrete in tension is adhered to, this form becomes poor practice. The closer the ribs are spaced, the less objectionable does such a form become.

Melan Arch with Metal Lath.—An arch very similar to the Melan system, but with the addition of sheets of metal lath laid between the curved ribs, has been used successfully in warehouse construction at a very low cost.* Cast-iron clips are suspended regularly, about 14 ins. centres, from the bottom flanges of the beams, and between these clips light curved channels are sprung. The channels used weigh about one pound per foot. On the tops of these channels are laid sheets of metal lath for use as a templet, or centering, on which to construct the concrete arch, and also to provide a better bond for the hard-mortar ceiling finish. A wrapping of metal lath passes around the beam-flanges. This is made in separate sheets, which are turned in under the concrete arch, when the concrete is built on the flanges of the beams, in order to prevent any possible displacement. The concrete is composed of 1 part Louisville cement to 4 parts cinders. The finished floor is made of 2 parts crushed granite to 1 part Portland cement, to give a hard wearing surface. The weight of this construction averages 35 lbs. per square foot.

Although very similar in form to the Melan arch, this great difference is to be noted. While the Melan system depends wholly upon the curved metal ribs for its load-carrying capacity, this other form depends entirely upon the action of the concrete as an arch. The light channels and the metal lath are used to provide a suitable centering only.

Roebling Arch Systems.—The most widely known floor made by the John A. Roebling's Sons Company, is shown in Fig. 60. It consists of a steel-ribbed wire-cloth centering and a cinder concrete arch. The wire centering is permanent, and is made of wire-cloth stiffened with $\frac{3}{8}$-in. to $\frac{1}{2}$-in. diam. steel rods woven into the cloth about 9 ins. centres. This cen-

* See "Fireproofing of Warehouses," by F. B. Abbot. *Journal Western Soc. Engs.*, April, 1898.

tering is sprung between the supporting I-beams in the form
of an arch, with the ends of the steel rods abutting into the
seats formed by the webs and lower flanges of the I-beams.
Longitudinally the sheets are well lapped and securely laced.

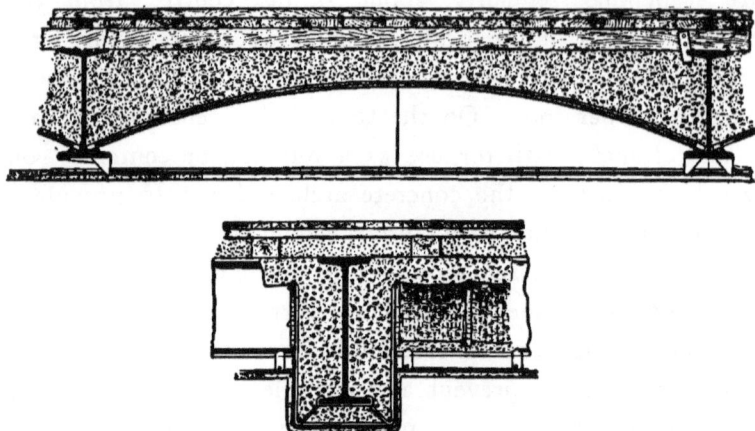

Fig. 60.—Roebling Concrete Arch, with Suspended Ceiling.

A cinder concrete is then filled in up to the tops of the beams,
giving a thickness of not less than 3 ins. at the crown of the
arch. The concrete is generally made of 1 part Portland
cement, $2\frac{1}{2}$ parts sharp sand, and 6 parts clean anthracite coal-
cinders.

The suspended ceiling is supported by flat bars, spaced
16 ins. centres, which are attached to the under sides of the
I-beams by means of patent clamps. In all spans over
3 ft. 6 ins., a $\frac{5}{16}$-in. steel rod is laid over and laced to the
supporting bars, and to this rod wire hangers are attached,
dropped from the crown of the floor arch at intervals not
exceeding 32 ins.   Wire lathing is then laced to the under side
of the supporting bars. This lathing has $\frac{1}{4}$-in. steel rods
woven in every $7\frac{1}{2}$ ins., and the sheets are so placed that
these wires cross the supporting bars at right angles, all
intersections being laced with galvanized wire.

This system is employed in office and public buildings where a perfectly flat ceiling is desired. The ceiling construction forms the only protection to the flanges of the I-beams. Tests have shown (see Chapter III) that the fire-resisting qualities of such a ceiling, with a one-inch space under the beams, will protect the beams from moderate fire a sufficiently long time to prevent their serious injury, although the ceiling is generally ruined. The ceiling may be placed any distance below the beams, and discoloration of the plaster, due to contact with the beams, may thus be avoided.

FIG. 61.—Roebling Concrete Arch, Exposed Soffit.

For factories, warehouses, breweries, etc., where a flat ceiling is not considered a requisite, a form is used as shown in Fig. 61. The construction of the floor is essentially the same as in the previous type, except that the curved soffit is left exposed and the lower flanges of the beams are encased by wire lathing, bent to a semicircular form and fastened to the wire centering at the sides, so as to leave at least 2 ins. of solid concrete around the beams at all points.

For shorter spans between beams, and where piping is not required over the beams, the floor-sleepers may be made to run parallel to the beams but depressed between them, so that the rough flooring may rest directly on the tops of the beams. The arch is made flatter, and the total floor-depth is reduced by the thickness of the sleepers.

Another type combines a suspended ceiling with the construction shown in Fig. 60. This gives a double protection to the beam-flanges, due to the ceiling itself and also to the concrete filling. Flat iron bars about $1\frac{1}{2}$ ins. by $\frac{3}{16}$ ins., spaced

16 ins. centres, are hung from the beam-flanges, and the lower surface is covered with No. 20 wire lathing, stiffened by steel wires. In spans of over 3 ft. 6 ins. the ceiling is supported at the centre, as before described.

The wire centerings for the above types are made of the proper size and form at the factory, so that they are rapidly and easily placed in position at the building. They form a valuable safeguard against the falling of the workmen, as they are sufficiently strong to sustain considerable loads in themselves.

For these floors, when the concrete is levelled up to the tops of the beams, the spans are usually made as follows:

Depth of beams or thickness of concrete at haunches.	Maximum span.	Thickness of crown at centre of arch.	Weight per sq. ft. including concrete and wire centering.
8 ins.	4 ft. 0 ins.	3 ins.	33 lbs.
9 ins.	4 ft. 6 ins.	3 ins.	34 lbs.
10 ins.	5 ft. 0 ins.	3 ins.	36 lbs.
12 ins.	6 ft. 0 ins.	3 ins.	41 lbs.
15 ins.	7 ft. 6 ins.	3 ins.	47 lbs.

Advantages.—Adequate tests have shown these floors to possess remarkable strength, and the fire-tests described in Chapter IV indicate generally good fireproofing qualities. They are easily and rapidly placed without the aid of wooden centering, and their construction is readily suited to all irregular conditions of framing.

Disadvantages.—In the form illustrated by Fig. 60, the lower flanges of the beams are protected by the ceiling construction only. This protection cannot be relied upon in severe fires, and any system which depends upon plaster only for any portion of the insulation cannot be considered first-class fireproofing. The large amount of water used in the mixing of the concrete, forms, in common with the other concrete systems, a disadvantage as regards the length of time necessary to thoroughly dry out the building before the interior finish can be placed.

Expanded Metal Co.'s Floors.—These floors are controlled and constructed by the various companies acting as licensees from the parent Expanded Metal Co., which first introduced the metal lath known as expanded metal. The systems rely upon a combination of concrete and metal, the latter being generally expanded metal, either with or without other reinforcement.

In early attempts at concrete floor-construction a rich cement-mortar was used at the bottom of the slab or plate, the idea being to give the material a greater tensile strength at the lower fibres. As this did not properly attain the object sought, metal was introduced in the lower part of the concrete slab, the cohesion between steel and concrete being practically ten times the tensile strength of concrete. Expanded metal thus embedded in the lower layer of a concrete slab has been developed into a great variety of forms, which have had a wide application in all sections of this country.

On wide spans of from 8 to 16 feet, arched channels, spaced about 4 ft. centres, are sprung from girder to girder to reinforce and stiffen the floor-plate. Concrete ribs are then built up on these channels to the level of the tops of the girders, and sheets of expanded metal are laid horizontally over the tops of the beams and the concrete ribs, to receive the concrete floor-plate, which is made 3 ins. or 4 ins. thick according to the loads to be carried. (See Fig. 62.) The channels usually employed vary in size from 5 ins. to 8 ins., weighing from $6\frac{1}{2}$ to $11\frac{1}{4}$ lbs. per foot. Wooden screeds are embedded in the concrete plate to receive the finished flooring. This form is usually known as the "Golding System," named after Mr. Golding, the inventor of expanded metal.

When the spans are less than 8 ft., the arched channels are dispensed with, and the system then depends upon the expanded metal for strength. (See Fig 63.) This form has been largely used in warehouses, hotels, and office buildings. It is used either with or without a suspended ceiling.   In case the latter is desired, this is accomplished by attaching small chan-

FIG. 62.—Expanded-metal Concrete Floor—Arched System.

nels or angle-irons, spaced 12 to 16 ins. on centres, to the bottoms of the beams by means of malleable-iron clips.   Expanded

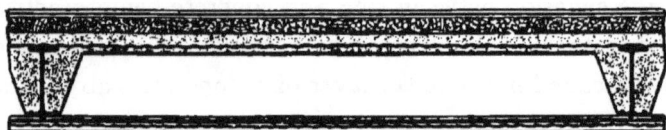

FIG. 63.—Expanded-metal Concrete Floor with Suspended Ceiling.

metal lathing is attached to these channels or angles with No. 19 annealed wire.

Another method, corresponding to the Roebling system, employs sheets of expanded metal sprung between the beams as a permanent arched centering, upon which a concrete arch is built. This form is well adapted to heavy loads, but it should be employed only when the span is small enough to allow a central height of $\frac{1}{8}$ to $\frac{1}{9}$ of the span.

Fig. 64 illustrates this form as used in the U. S. Public Building at San Francisco, Cal.   The floor-arches were $3\frac{1}{2}$ ins. thick at the crown, and were composed of concrete made of 1 part Portland cement, 2 parts sand, and 5 parts furnace-clinkers. In the soffit of the arch was embedded either No. 18 expanded metal, 3-in. mesh, or No 19 wire cloth with $\frac{7}{16}$-in. rods woven in every 9 ins.   The suspended ceiling was made of 1-in. by 1-in. tees, spaced 16 ins. centres, secured to the beam-flanges by clips. Expanded metal or wire lath was laced to the tees to receive the plaster finish. All girders projecting below the ceiling-line were encased in terra-cotta shoe-blocks. The finished floor was made of tile or terrazo.

When a fireproof floor is desired to produce a level ceiling at a low cost, an expanded-metal floor has often been con-

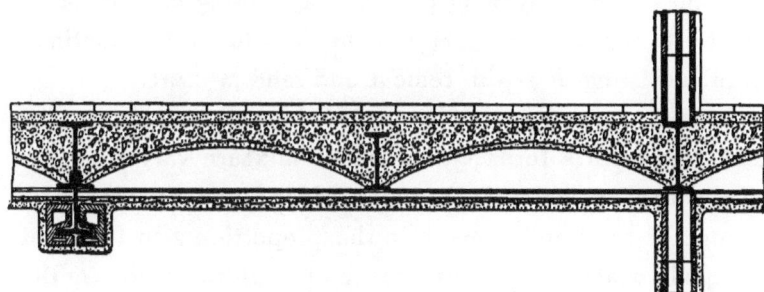

FIG. 64.—Concrete Floor-arches in U. S. Public Building, San Francisco.

structed by laying sheets of expanded metal upon the lower flanges of the supporting I-beams. A 3-in. concrete plate is then placed and tamped, and a lighter cinder concrete is filled up to the tops of the beams and between the floor-screeds. This system is usually employed in apartment houses, where the beams may be made 5 ins. to 7 ins. deep, spaced not over 4 ft. centres. This construction is not adapted to heavy loads.

Method of Setting.—In all of these forms except the arched system, a wooden centering is suspended at the proper level to receive the horizontal sheets of expanded metal, and to sustain the concrete while it is tamped, and until it has set. In the Golding system the centering is constructed with open forms to receive the arched ribs which project below the flat surface, and with bevelled sides to form the concrete protection around the beams. Sheets of water-proof paper are always placed upon the centers to keep the sand and cement from running through cracks or holes in the centers. This gives a layer of cement-mortar at the bottom of the slab. The centers should remain in place at least five days in good weather. The floor may be used 24 hours after setting.

The expanded metal is either 3-in. No. 10 gauge or 6-in. No. 6 gauge, placed lengthwise across the beams in sheets 8 ft. by 5 ft. The concrete is then spread and tamped, so that the metal is embedded in the lower inch of the plate. Nailing-strips may be employed, or the wooden flooring may be nailed directly to the concrete. A cement floor finish is sometimes given consisting of a 1-in. cement and sand mixture.

The concrete is generally made of 1 part cement, 2 parts sand, and 6 parts furnace-cinders, the mixture varying somewhat with the character of the sand and the cinders. The amount of sand and cement, in the proportion 2 to 1, should be equal to about $1\frac{1}{4}$ times the volume of the voids in the cinders. This volume may be ascertained by using water to fill the voids. The cinder concrete will weigh about 84 lbs. per cubic foot, and the cinder filling between the screeds about 60 lbs. per cubic foot. The thickness of the slab or plate of concrete also varies according to the spans and load. A 3-in. slab is usually sufficient for all ordinary floors.

Advantages. — The systems are all easily adaptable to irregular framing, and they are comparatively light in weight · and cheap in cost

Disadvantages.—Except in the arched form, the principal reliance of tensile strength is placed upon the sheets of expanded metal embedded in the lower portion of the floor-plate. Unless these thin sheets of metal are encased in cement or cement-mortar, the possible corrosive action of the cinders employed raises serious doubts as to the ultimate life of the expanded metal. The large amount of water which drips through the successive stories during construction also forms a serious drawback, in that a long time is required to thoroughly dry out the building so that the interior wood trim may be placed. The type usually recommended by these companies for office buildings or warehouses is as shown in Fig. 63. When used with a suspended ceiling, the lower flanges of the beams are protected by the plastered ceiling only, which is not sufficient for unquestioned protection.

Mackolite Floor.—This floor differs materially from most of the constructions in general use, in that the material is designed to act as a beam or lintel instead of as an arch. The floor, as shown in Fig. 65, consists of fireproof slabs or blocks

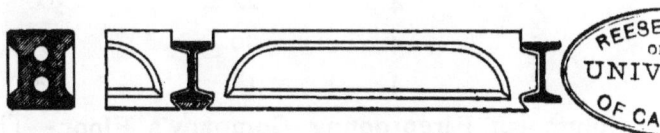

FIG. 65.—Mackolite Floor-construction.

whose ends rest upon the supporting floor-beams. These blocks are made of "mackolite," which is calcined and ground gypsum, mixed with water and chemicals. The mixture is moulded in forms, being left for about $1\frac{1}{2}$ hours to set, after which it is kiln-dried for four days. These blocks are made up to a maximum length of 5 ft., the floor-lintels being 10 ins. deep and 9 ins. wide. Lintels 7 ins. deep are used for roof-construction. The section of the block is rectangular at the ends, but the diagonally opposite vertical corners are rounded off, to permit the placing of the block between the floor-beams. Sections of flange-protection tiles 4 ft. long and $1\frac{1}{4}$ ins. thick are held under the beam-flanges by the dovetailed projections from the main blocks. No centering is required for erection, and as there is no end thrust no tie-rods are necessary. Cinder concrete with nailing-strips and finished floor is employed as usual, while the ceiling-soffit is plastered directly on the blocks.

The weight of this system is given by the manufacturers as 26 lbs. per sq. ft. for blocks 10 ins. deep, and 22 lbs. per sq. ft. for the 7-in. roof-blocks.

The following table gives the results of some tests made

on these blocks, the figures being the averages for ten tests in each case : *

Depth of Blocks, inches.	Span, feet.	Weight of Blocks per Square Foot.	Load per Square Foot, corresponding to Centre Breaking-load.
10	4	26	575
10	4	30	825
10	5	26	460
10	5	30	600
7	4	22	500
7	4	20	450
7	4	18	400

Columbian Fireproofing Company's Floor.—This system of fireproof floor-construction, patented and controlled by the Columbian Fireproofing Company, consists of a combination of rolled-steel bars and concrete. The bars, which are rolled with a vertical web and horizontal flanges or ribs, are usually suspended from and attached to the floor-beams by means of steel stirrups. These are cut or perforated to the exact shape of the bar used, and are made to straddle the top flanges of the I-beams. The bars and stirrups are then sur-

FIG. 66.—Columbian Concrete Floor—Panelled Construction.

rounded by and embedded in cement concrete, which is deposited on a wooden centering hung beneath the bars.

Two general forms of construction are used : the "panelled," in which the floor alone is constructed, and the "flat-

---

* See *Engineering News,* May 26, 1898.

ceiling construction," in which separate slabs are employed for both floor and ceiling.

The panelled construction is shown in Fig. 66. The lower portions of the beams, projecting below the ceiling-line, are either encased in solid concrete, as shown in Fig. 66, or they may be protected by means of concrete slabs, placed so as to leave an air-space surrounding the steel. (See Fig. 67.) The latter form, however, is not recommended by the Columbian Fireproofing Co. In both cases the bottom flanges of the beams are first covered by means of concrete slabs which are made at the building. These are moulded in forms to suit the different widths of beam-flanges.

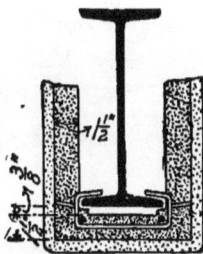

FIG. 67. — Beam-casing — Columbian System.

Iron straps or clamps made of pieces of No. 16 iron, ¾ ins. wide, are embedded in the mixture during the moulding of the blocks, and pieces of No. 10 galvanized wire are also embedded lengthwise, as shown in Fig. 67; so that in case the concrete is broken away or cut around the bands, the embedded wires will still hold the clamps in place and prevent the falling of the blocks. The blocks are made about 12 ins. long. The clamps have projecting ends which are bent around the beam-flanges, and are themselves, in turn, protected from fire by means of the solid concrete, or concrete slabs, used to encase the beam-web. A ⅜-in. air-space is left below the beam in either case.

The spans employed vary from 4 ft. to 8 ft., 6 ft. to 7 ft. being generally the most economical spacing for the beams. This panelled construction is suitable for use in warehouses or mercantile buildings in which a panelled ceiling is not objectionable. In many cases the concrete is left exposed without plaster finish. The imperfections in the surface are gone over with mortar, leaving a finish equal to a well-pointed brick wall. Whitewash is often applied, as in the new buildings lately

erected for the Boston Wharf Co., in Boston, where some
470,000 square feet of floor-area were treated in this manner.
If plaster is used, two coats only need be applied.

The flat ceiling construction, shown in Fig. 68, consists of
a floor-plate as in the previous form, with the addition of
lighter bars, resting on the lower flanges of the beams, on

FIG. 68.—Columbian Concrete Floor—Flat Ceiling Construction.

which bars a concrete ceiling-slab is cast. An air-space is left
between the two plates. The beam-webs are either left exposed or are encased in concrete, but the exposed webs are
practically as well protected by the ceiling construction as are
the encased webs in the panelled form.

This form is not generally used for spans greater than 7 ft.
For light loads, as in apartment houses, etc., this system is
often used without the employment of stirrups. In this case
the bars pass directly over and rest on the floor-beams. If
different-sized beams occur, flush on the bottoms, inverted
stirrups can be used, so as to allow the bars to run over the
lightest beams.

The ribbed bars used in the floor construction are made in
three depths: $1\frac{1}{2}$-inch, 2-inch light and 2-inch heavy, $2\frac{1}{2}$-inch
light and $2\frac{1}{2}$-inch heavy. The 2-inch and $2\frac{1}{2}$-inch bars have
two horizontal ribs each, while the $1\frac{1}{2}$-inch bars have but one
rib. The shapes of the bars and stirrups are shown in Fig. 69.
The spacing and depths of the bars used are varied to suit the

requirements as to load and span. The 1½-inch bars are employed in residences, apartment houses, etc., the 2-inch sections in office buildings or hotels, and the 2½-inch sections in warehouses, storage, and manufacturing buildings. The 1-inch bars are used in the ceiling construction and in short spans where the floor-bars pass over the supporting beams. The

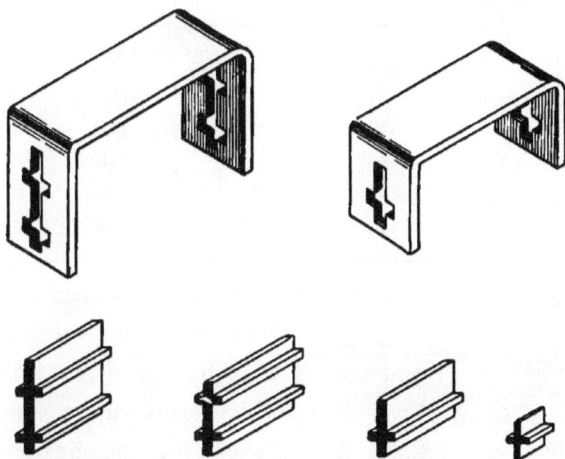

FIG. 69.—Stirrups and Bars used in Columbian System.

bars, for all sizes, are generally spaced from 20 ins. to 24 ins. centres, and are cut to the lengths required at the building site. The 1-inch ceiling-bars may also be bent at the site to form domes or other special features.

In buildings of a heavy character the concrete is composed of Portland cement and stone, or furnace slag, where the latter can be obtained. The mixture generally employed is 1 part Portland cement, 2½ parts sand, and 5 parts broken stone. Cinder concrete is often used in the ceiling-slabs, or for floors in lighter structures where no great loads are to be carried. This mixture is made of 1 part Portland cement, 5 parts cinders, and enough sand to properly fill the voids— usually 2½ parts to the above proportions of cement and ;inders.

The floor is finished by means of nailing-strips embedded in cinder concrete filling, or in a concrete made of ¼-inch stone screenings.

The weights per square foot for these constructions, for the bars and concrete only, are as follows: ·

Size of Bars. Inches.	Thickness of Floor. Inches.	Panelled Construction. Solid Casing.		Flat Ceiling Construction.	
		Stone Concrete. Pounds.	Cinder Concrete. Pounds.	Stone Concrete. Pounds.	Cinder Concrete. Pounds.
1	2¼	42	26½	48½	37
1½	2¾	42	26½	48½	37
2	3¼	46	29	54½	40½
2½	3¾	54	35½	59½	43½

Method of Setting.—For the panelled construction, wooden centers are supported by means of 2-in. by ⅜-in. iron shoes, which rest upon the lower flanges of the beams. These shoes carry longitudinal plank (forming the side of the web casing), which in turn carry 3-in. by 4-in. cross-pieces spaced 2 ft. centres. Upon these rest

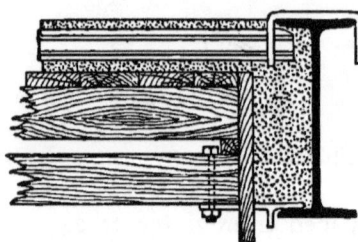

FIG. 70.—Floor Centers—Columbian System.

the ⅞-in. planking which receives the concrete floor-plate. (See Fig. 70.)

For the flat ceiling construction the ceiling-plate is first made on a wooden centering, which is hung from the lower flanges of the beams by means of iron hangers. Then, before this centering is removed, the centers for the floor-plate are built directly upon the ceiling-slab. When both have been built, and the concrete has set, the ceiling center is first removed. The floor centering is then taken out of the intermediate space through a rectangular opening, about 18 ins. by 22 ins., which is left in the ceiling, one opening to each bay. The centers for the floor-plate are supported on wood wedges (resting on the ceiling-plate), which are first removed by using a hook. The rest of the floor centering is then removed piece by piece, the sections used being of such lengths as will permit their withdrawal through the opening. For 12-in. beams, 4-ft. lengths are used as a maximum; for 15-in. beams, 8-ft. sections can be employed. After the floor centering has been entirely removed, the opening is closed up by a concrete slab having beveled edges.

The centers should remain in place from 2 to 5 days, according to the weather and quality of cement used. The floor can be walked on without indentation in from 1 to 5 hours after setting.

No tie-rods are used in this system, nor are wall-channels necessary where masonry walls occur. A horizontal chase 2 ins. deep, or the depth of one brick, is left by the masons, and the floor-bars and concrete then run into and fill this space. This avoids ceiling-cracks along the wall-lines.

Advantages.—The floor can readily be adapted to irregular framing, and can be easily and quickly set. Six thousand square feet have been laid in a single day, with service of one elevator. Great strength is claimed, and superior fire-resisting qualities. It is certainly probable that the stone employed in the concrete mixture is preferable to the very poor quality of cinders generally used in most cinder concrete constructions, and the requirements of protection for the steel work against corrosion are fully met by the stone concrete used.

Disadvantages.—The concrete used is generally very wet, and a long time is required to dry out the building thoroughly. In setting the floors the lower stories are almost always placed first, so that, in working up, the water drips through to all floors below. This makes it impossible to start wood finish in the lower stories until all floors and roof are completed and well dried out.

The Metropolitan Floor.—This construction, formerly known as the " Manhattan " system, is manufactured by the Metropolitan Fireproofing Co., of Trenton, N. J.   The floor-plate is similar in form to the Columbian, except that wire sus-pension-cables are used for the supporting metal members, in place of ribbed bars.

Cables, each composed of two No. 12 galvanized wires, twisted, are first laid across the tops of the floor-beams and secured to the walls by means of anchors ; or, where the cables terminate at a beam, they are secured to the top flange by means of strong hooks, 3 ins. long, made of $\frac{1}{4}$-inch iron. These cables are spaced from $\frac{7}{8}$ ins. to $1\frac{1}{2}$ ins. centres, according to the spans and loads, and are laid parallel.   Lengths of $\frac{7}{8}$-in. round iron rods are laid on the cables midway between the beams, so as to cause the cables to deflect uniformly 3 ins. below the tops of the I-beams in a 6-ft. arch.   (See Fig. 71.)

FIG. 71.—Metropolitan Floor—Panelled Construction.

Forms or centers are then placed between the floor-beams and 1 inch below the iron rods.   A composition formed of about 1 part plaster of Paris by bulk to 2 parts of spruce or hemlock planer shavings, with sufficient water to mix thor-oughly, is then poured in place, tamped, and brought to a level $\frac{1}{2}$ in. above the tops of the floor-beams.   This forms a floor-plate $4\frac{1}{2}$ ins. thick, ready for the laying of wooden sleepers, filling, and finished floor.   The portions of the beams project-ing below the ceiling-line are protected by means of a two-inch thickness of the composition, poured at the same time the arch is made, into forms left in the centers.   The bottom flanges of the beams are wrapped with wire netting before the composition is poured, so as to prevent the plaster from cracking away from the beam.

When a flat ceiling is desired, $\frac{3}{4}$-in. by $\frac{1}{4}$-in. flat bars, spaced about 16 ins. centres, are suspended from the lower flanges of the I-beams by means of clips, so that the tops of these flats are about $1\frac{1}{4}$ ins. below the bottom flanges of the beams. No. 22 wire lathing is then fastened to the under side of the flats to receive the plastering. (See Fig. 72.)

For the flat ceiling construction, the beams projecting below the floor-plate are covered with blocks of the same composition as the arch, 2 ins. in thickness, and securely fastened in place by means of plaster of Paris backing and joints. These beam-protection blocks are made in wooden forms at the building site. Four blocks are used, two for the flange and two for the web, as shown in previous figure.

Method of Setting.—The centers are made of $\frac{7}{8}$-in. rough boards, supported on 3-in. by 4-in. stringers which are suspended below the beams by means of wire loops. Where a

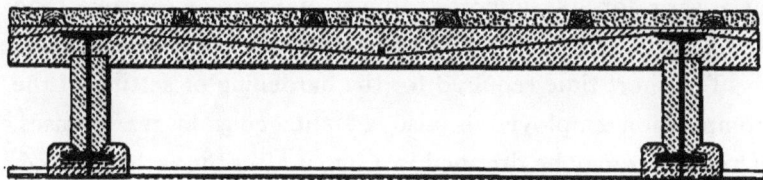

FIG. 72.—Metropolitan Floor—Flat Ceiling Construction.

suspended ceiling is not used, the centers are covered with a rubber mat or sheet before the mixture is poured in place. Upon the removal of the centers this mat is stripped off, leaving the composition surface smooth and ready for the finishing coat of plaster, without requiring rough plastering.

Nailing-strips with cinder concrete filling are usually laid on the floor-plate to receive the finished floor, the cinder concrete being laid as dry as practicable. The nailing-strips, or screeds, cannot be nailed directly to the floor-plate, but they are sometimes run at right angles to the beams, in which case they are toe-nailed to the top flanges.

Spans over 8 ft. are not to be recommended. A 6-ft. span is considered the most economical, and the most satisfactory in construction. In this case the wires are laid about $1\frac{1}{4}$ ins. apart. In cases where the spans and loads require only light beams, say 5-in., the floor-plate is made solid for the full depth of the beam. This detail is often employed in apartment houses.

No tie-rods are required for this system.

Weight.—The composition employed weighs about 60 lbs. per cubic foot. A $4\frac{1}{2}$-in. floor-plate weighs 20 lbs. per sq. ft.

Advantages.—The great advantage of the Metropolitan system lies in its lightness. It is, beyond question, the lightest form of fireproof floor having any extended use, but it is to be remembered that lightness in itself is no advantage (except as regards the foundation-loads in a heavy or high structure) unless accompanied by an attendant reduction in cost or efficiency. A floor system may reduce the quantity of steel frame necessary for its support, and yet be more expensive than some construction requiring a heavier frame.

The short time required for the hardening or setting of the composition employed is also advantageous in many cases. The centers may be dropped in 4 hours after the arch is placed, and in 15 minutes after the mixture is poured and smoothed the floor may be walked on without indentation.

No skilled labor is required, save a competent foreman. The mixture does not require placing with any great regularity, as a perfect bond is always secured between the surfaces, no matter how rapidly the material may set.

Disadvantages.—In this construction, black iron-wire cables were formerly used. These caused a discoloration due to rusting. Galvanized cables are now used to remedy this defect, and they are generally found to be satisfactory.

Trouble is liable to occur, however, from discolorations due to sap in the shavings employed, where the lumber from which the shavings are planed is not thoroughly kiln-dried. This produces a light yellow stain, which often strikes through the plaster, and causes great trouble and expense in renewing ceiling decorations. Portland cement is sometimes added to the mixture, which, in falling to the bottom of the plate, gives a cement finish to the ceiling, which helps to prevent discoloration.

Difficulty often results from the trying out of the water in cases where the floors are laid from the basement up, in cold weather. This is due to the exposure of the successive floors to the weather, and also to the moisture caused by mixing the composition over the finished floors. Water keeps dripping through to the floor below, and the centers also allow small quantities of moisture to drip. The preferable way is to start at the roof and work down. This saves all dripping, and the floors are always under cover if the roof is made water-tight.

Rapp System. — The several forms of this system are made up of sheet-iron tees laid on the lower flanges of the supporting I-beams, bricks laid between these T's, and cinder-concrete filling up to the tops of the beams. Three types are used—flat, segmental, and panelled.

FIG. 73.—Rapp Floor—Flat Construction.

The flat construction is as shown in Fig. 73. Steel T's, rolled cold from the flat, are laid horizontally across the beams, and rest on the bottom flanges. They are about $1\frac{1}{2}$ ins. $\times$ $1\frac{1}{2}$ ins. in section and are spaced. $8\frac{1}{2}$ ins. centres, being held in position and stiffened by steel spacing-ties, as shown in the illustration. Bricks are then laid between the T's and grouted, and the space up to the beam-tops is filled in with cinder concrete. The mixture usually employed is eight parts cinders to one part Portland cement. The lower beam-flanges are wrapped with wire lath. Nailing-strips, cinder filling and wood flooring give the usual finish for the floor, while the ceiling is plastered directly on the flat brick surface.

The segmental construction is essentially the same as the foregoing example, except that curved or arched T's are used in place of the flat ones. This form is used where the supporting beams are deep and where additional floor-strength is required. The arching develops increased strength, and diminishes the amount of concrete necessary. The bottom flanges of the beams are protected by wire lath or by plaster blocks. If a flat ceiling is desired, a hung ceiling of wire lath is suspended below. Another form of segmental arch is used in which the bricks are laid on edge, thus giving a four-inch arch, as required by the building laws of the city of New York.

In the panelled construction the T's are bent to a shape as shown in Fig. 74. The angle of the panel can be made as

FIG. 74.—Rapp Floor—Panelled Construction.

great or small as is desirable, up to a right angle approximately, the brick and concrete being placed as in previous cases. This form is used where it is desired to keep the weight of the construction down as low as possible. A suspended ceiling may be used as before.

The supporting beams are usually spaced from 4 ft. to 5 ft 6 ins. for the flat construction, and from 6 ft. to 9 ft. for the arched system. The arched type has been used up to 13 ft. span, by employing larger tees.

The weight of the tee-bars and brick, set in place, is 21 lbs. per sq. ft.

Advantages.—A great advantage and saving of cost in these systems lies in the fact that no centering is required. The tees are cut to measure and coated with asphalt paint at the factory, and can be rapidly placed in position when required. The tees have sufficient strength in themselves to permit walking on before the bricks and concrete are laid. The construction is also well adapted to irregular conditions and to roof, dormer, and dome construction.

Disadvantages.—The use of small sheet-iron tees, protected only by a coating of plaster, cannot be considered first-class fireproof construction. The system is undoubtedly a cheap one, and is being used to a considerable extent in apartment houses, hotels, and the like; but its adoption in any building intended to be fireproof and of first-class permanent construction is not to be recommended. The test by the New York Building Department shows that it is very weak under fire- and water-tests, and that a large amount of reconstruction would be necessary.

Roebling Flat Construction.—Besides the various arched systems before described, the John A. Roebling's Sons Company employs a form of flat floor construction, illustrated in Fig. 75, *A*. Flat steel bars are employed, set on edge and spaced 16 ins. centres, with a $\frac{1}{4}$ turn at the ends where the bars rest upon the I-beams; or, in longer spans where deeper beams are employed, the bars have a $\frac{1}{4}$ turn forming a hook to

fit over the upper flanges of the beams, as shown in Fig. 75, *B*.
Steel-rod spacers are inserted at intervals to separate and brace
the bars. Wire lathing with $\frac{1}{4}$-in. stiffening-rods woven in
every $7\frac{1}{2}$ ins. is then applied to the under side of the bars, the
stiffening-rods running crosswise under the bars, and laced to
them at every intersection. A cinder concrete plate is then

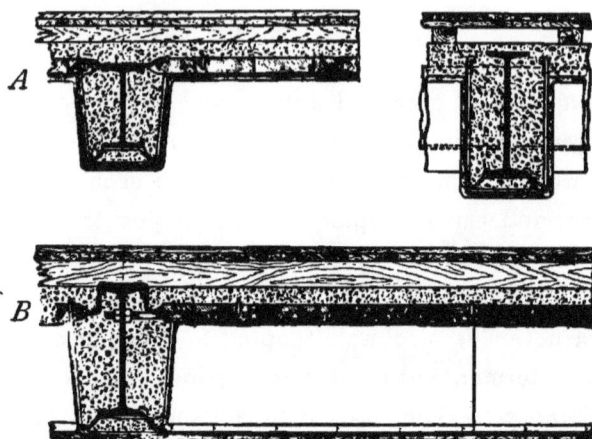

FIG. 75. Roebling Floor—Flat Construction.

formed on this flat metal centering $3\frac{1}{2}$ or 4 ins. thick from
bottom of slab to the under side of the nailing-strips, which are
usually employed with wood flooring.

A flat ceiling may be added, made either with rods
suspended from the beams by clamps, or by using flat bars set
on edge, as in Fig. 75, *B*, when the span is considerable. The
ceiling when plastered finishes $2\frac{1}{2}$ ins. below the under side of
the floor-beams.

The portions of the beams projecting below the floor-slab
are encased by stiffened wire lathing which is supported on
light iron furring built out so as to keep the lathing at least
2 ins. from the beam at all points. The space between the
beam and the wire lathing is then filled solid with concrete.

For floors with light loads, such as residences, apartment

houses, hospitals, etc., the upright floor-bars are sometimes supported on the lower flanges of the I-beams, and the floor-sleepers depressed, so that the total floor thickness consists of the beam depth plus the plastering below and the wooden floor-thickness above.   Bars 2 ins. $\times$ $\frac{1}{4}$ in. are used in this case, spaced 16 ins. centres for a span of 6 ft.

Fawcett System.—For the Fawcett ventilated fireproof floor a special patented form of terra-cotta lintel is used, the idea being to insulate completely from heat the lower flanges of the supporting beams by the interposition of an air-space between the metal-work and the terra-cotta.

Fig. 76 shows the terra-cotta lintels without concrete filling. These lintels are usually made of hard-burned terra-cotta.

FIG. 76.—Fawcett Floor Construction.

The width of the lintels is constant, 9 ins., in the direction parallel to the beams.   The height is also constant, or $5\frac{1}{2}$ ins. for all lengths.

For the purpose of erection, as allowing the pieces to be swung into position between the beams, the lintels are laid at such an angle with the beams as to make the shorter diagonal of each lintel come at right angles to the beams. Lintels for different spans have their ends cut at slightly different angles to give this result.

The lintels bear on and fit around the lower flanges of the supporting beams in such a manner as to leave an air-space of about $\frac{3}{4}$ ins. under the entire length of each beam-flange.   This air-space connects with the interior of each lintel, thus providing a connected air-space under all beams and through all lintels.   At intervals, usually in each bay, air-flues or thimbles may be built into the exterior walls, thus connecting the interior air-spaces with the outside atmosphere. These air-flues, or " ventilating-bricks " as they are called, are made the size of a single brick, with a grating at the inlet to prevent the admission of dirt, rubbish, or birds. Next to the exterior wall a "split" or half-lintel is placed, into which the ventilating-brick may lead.

In case of fire, the heated air in the lintels and under the beam-flanges is supposed to circulate freely enough to secure the proper insulation of the metal-work. In many cases where the ventilating-bricks are not used, the volume of confined air is considered to be sufficiently large to act as a cushion with enough elasticity to prevent the explosion of the confined and heated air. The ventilating-bricks are usually employed in buildings having brick walls, but where stone walls or skeleton construction is used, it is often difficult to make a suitable connection with the outside air.

After the lintels have been placed, a cinder concrete made of six parts coarse cinders to one part Atlas Portland cement is filled in to the height of about 2 ins. above the tops of the beams. Nailing-strips are embedded in the concrete; these may either be placed directly upon the tops of the beams, or, as is more usual, they may run at right angles to the beams.

In erection the lintels are set without mortar. No reliance is placed upon the lintels for ultimate strength, as they are employed as centers only to receive the load-bearing concrete, and to provide the air-space before mentioned. They are readily swung into position, until bearing firmly against the beam-web and flanges, but if an irregular bearing is secured, the brittleness of the hard-burned material may, under a considerable loading, cause the lintels to crack at their supports. The concrete is then rammed in place up to the tops of the I-beams, ready to receive the nailing-strips. The lintels and concrete filling may be placed very rapidly, as is evidenced by the laying of an area of 10,000 square feet in a single day.

For the supporting metal-work small I-beams are employed, usually 4-in. to 7-in., spaced 2 ft. 6 ins. centres where possible, and ranging in span from 10 ft. to 22 ft. These are carried by girders running between the columns or other supports.

The small I-beams which carry the lintels are not considered by the Fawcett Company as being governed by the requirements for floor-beams specified in municipal building laws. An entire bay, or the area within the lines of the supporting girders, is considered as composite Fawcett Construction, much as a system of concrete and twisted rods would be treated as special construction. The concrete is relied upon to so stiffen these small beams as to warrant the use of higher extreme fibre-strains than are ordinarily employed, and to disregard largely the usual limits set for permissible deflection. The extreme fibre-strains employed vary from 16,000 lbs. to 24,000 lbs. per sq. in., according to the class of building and assumed live load. Where municipal requirements specify a live-load that is considered excessive, the 24,000-lb. fibre-strain is used. This is apt to result in an excessive vibration of the floor-system, especially during erection or until the concrete has become well set. The special lintels weigh, on an average, about 14 lbs. per sq. ft.

Advantages.—The advantages of the Fawcett System lie in the shallowness of the floor (which may be important in cities where the building height is limited), in the absence of tie-rods, and in the facility of setting. The shallowness, however, loses its value where the girders cannot be placed in partitions.

Disadvantages result from the necessary redesign of the steelwork to suit the standard lintels, and from the cost and difficulty caused by long spans, in which case the lintels retain the same shallow depth as in short spans, while the beams must be materially increased in depth, and still kept at a uniform distance of 2 ft. 6 ins. centres.   This greatly adds to the concrete filling and to the cost of the steelwork.

From the standpoint of good construction the shallowness of this system constitutes a weakness.   The employment of any system depending upon shallow beams is not to be recommended, on account of the attendant vibration both during construction and after completion.   The hard-burned terra-cotta lintels also are very liable to suffer breakage under shock; and even during erection, when the tiles do not receive an even bearing at the ends, they crack easily and necessitate constant reconstruction.

The fire-test by the New York Building Department (see Chapter IV) showed conclusively that the employment of hard-tile lintels, as used in this system, cannot be considered a satisfactory fireproof construction.

Other Forms of Fireproof Floors.—In addition to the types described above, many other forms of fireproof floors have been used, some quite extensively and others in but isolated cases.   Several systems which have not been described in the previous pages possess more than ordinary merit, but their peculiar construction necessarily limits their application.   This is true of the Poulson, or " Hecla " floor, and the Guastavino dome construction.   Both of these methods are admirably suited for use in public buildings, libraries, etc., where groined arches or domes are in keeping with the architectural effect desired, but they are not adapted to use in commercial or office buildings, or hotels, or apartment houses.   In the Guastavino form, also, the walls or piers must be of a strength sufficient to resist the thrusts induced by the dome-construction employed.

The Lee hollow-tile and cable-rod floor, combining terra-cotta blocks with suspension cables for use in long spans, has not had a general application, nor is it likely to have such in ordinary practice.

Other forms of fireproof floors have largely partaken of the nature of curiosities only, and have soon died a natural death after being patented.

Comparison in Weights.—As influencing the amount of the supporting steelwork required, the weight per square foot of the floor-system employed becomes a matter of consideration from an economical standpoint. Comparative weights have been calculated on the following assumptions.

The floor-system in each case has been designed for a bay or panel 16 ft. square, and it has been assumed that the columns supporting this panel are connected by girders in one direction, and by the floor-beams in the other direction. The weights of the floor-beams per square foot then include all of the beams wholly within the bay, and one beam running between the columns. The girder-load includes one girder only.

	Terra-cotta. End-Construction. See Fig. 35.	Roebling Floor. Arch Construction. See Fig. 60.	Columbian Floor. Flat-ceiling Construction. See Fig. 68.	Expanded Metal Co.'s. No. 3 Construction. See Fig. 63.	Fawcett Floor. See Fig. 76.	Rapp, Floor. Panelled Construction. See Fig. 74.	Metropolitan Floor. See Fig. 72.
Floorbeams	6.0	6.0	6.0	4.0	6.5	6.0	6.0
Girders.................	3.5	3.5	4.0	3.5	3.5	3.5	3.5
Arch or slab	39.0	41.0	54.5	30.0	14.0	21.0	25.0
Concrete levelling........	7.0	......	......	......	30.0	21.0	......
Concrete filling..........	21.0	21.0	21.0	21.0	21.0	21.0	21.0
Screeds, 3-in. by 4-in., 16 ins. centres..........	1.5	1.5	1.5	1.5	1.5	1.5	1.5
Flooring, ⅞-in. maple....	4.0	4.0	4.0	4.0	4.0	4.0	4.0
Plaster on ceiling	5.0	......	5.0	5.0	5.0	5.0	5.0
Ceiling construction and plaster...................	......	10.0	......	2.0	......	10.0	2.0
Wires and centre rods.....	......	......	......	......	......	......	0.5
Total dead-load	87.0	87.0	96.0	71.0	85.5	93.0	68.5
Live-load	150.0	150.0	150.0	150.0	150.0	150.0	150.0
Total load per sq. ft......	237.0	237.0	246.0	221.0	235.5	243.0	218.5

The live-load is taken at 150 lbs. per square foot, and the beams and girders are calculated for their full dead- and live-loads, on the basis of a 16,000-lb. fibre strain.

The cinder concrete filling between the beams is assumed to weigh 84 lbs. per cubic foot in each case, and this has been taken for a thickness of 3 ins., allowing for 3-in. by 4-in. spruce screeds, 16-ins. centres. The flooring is assumed to be ⅞-in. maple. A suspended ceiling has been included for such forms as do not provide a flush ceiling construction.

Choice of Concrete or Composition Floor. — In the descriptions which have been given of the many concrete and composition floors now in use, it will be noticed that little or nothing has been said about the strength of the various systems under test-loads. A great quantity of data is available as showing the strengths of these floors under many varieties of conditions, some of which are quite reliable, while a large percentage are prejudiced or misleading. Great emphasis, also, is laid upon these test-loads by the fire-proofing companies in their catalogues. The selection, however, of any one system will almost certainly be from considerations other than mere strength under test conditions.

For municipal regulations, the Boston building law may be cited as even more severe than the usual requirements:

" The least capacity per superficial square foot, exclusive of materials, shall be: For floors of dwellings, fifty pounds; for office floors, one hundred pounds; for floors of public buildings, one hundred and fifty pounds; for store floors, floors of warehouses, and mercantile buildings of like character, drill-rooms and riding-schools, at least two hundred and fifty pounds." *

Under all ordinary conditions, almost any of the floor systems which have been described will develop a reasonable

---

* Boston Building Laws, § 55.

factor of safety under these loads. Failure under loads, other than extreme test-loads, has very seldom occurred in which the floor, as a system, was at fault. The failure can almost always be traced to careless laying or to poor workmanship at some vital point.

The selection of a system, whether from the many concrete types or from terra-cotta on the one hand and concrete or composition on the other, will depend almost wholly upon practical considerations. The author does not believe, as do many who have expressed a decided preference for one particular type, that the question of strength is the one important consideration.

As affecting a practical decision, the item of cost will naturally appeal first to the investor, but the *first cost* cannot represent the ultimate value under the slow test of years or the unexpected test of fire. The action of the system contemplated for use must be allowed for under these conditions, and when given due consideration, the *prima facie* evidence may be wholly reversed.

In the six floors other than terra-cotta for which calculated weights per square foot were made, " selection by rejection " may be made as follows:

Rapp Floor.—This can in nowise be counted as satisfactory fireproofing, for even though the arched system be used, and it be admitted, supposedly, that the concrete filling would constitute an arch sufficient to carry the loads after the inevitable destruction of the sheet-iron tees by fire, the resulting damage must still be great through the destruction of the suspended ceiling and the dropping of the bricks between the tees. The latter fact also constitutes a menace to those who must work under the floors in time of fire. The beam-flange protection relies wholly upon expanded metal and plaster. The New York Building Department test would seem conclusive as to the inadequacy of this system.

Fawcett Floor.—The thinness of this system, and its consequent deflections, combined with the poor showing made under fire- and water-tests, indicate that it is far inferior as a fireproofing method to many floors now in use. The hard-tile lintels continue to be employed in spite of the many failures of this material in both tests and actual fires. The New York tests showed that the hard lintels crack easily under the action of heat and water, and that many of them fall and expose the concrete filling. This means great difficulty and expense in reconstruction.

Metropolitan Floor.—For fire resistance this detail is very acceptable, though tests prove that the level-ceiling construction is thoroughly destroyed by fire, and the action of water washes away exposed portions of the slabs and renders the material soft at the surface. The nature of the material, however, makes reconstruction an easy matter.

Other practical objections of often serious moment are, however, to be found in the combination of plaster of Paris and wood shavings employed. In setting, the slabs will generally harden at the surface first, which sometimes causes the ceiling surface to tear away from the mass. Also the complete saturation of the wood shavings with moisture causes the mass to retain moisture for a very long period, as the gypsum will first set at the surface, thus retarding the evaporation of the water from the inside of the mass. This is slowly accomplished, and often with accompanying stains resulting from the use of poorly dried shavings, which are liable to cause great injury to ceiling decorations or tinting.

Previous to the introduction of galvanized cables, this system was open to severe criticism on account of the corrosive effect which gypsum in any form has upon iron or steel. This defect still exists as far as the beams are concerned, unless the painting is considered sufficient to prevent corrosive action.

Expanded Metal Floors.—The flat system, tested in the New York experiments, did not make a very good showing. The top of the floor was quite hot, and the tie-rods and beam-flanges lost nearly all of their protection. But the portions of the concrete slab which were washed away could be easily replaced in reconstruction, as could the suspended ceiling.

It would seem, however, that these flat systems, which depend so largely upon sheets of expanded metal for strength, could hardly be considered good engineering practice for permanent construction. The author believes that any design which depends upon the strength of No. 10 steel embedded in cinder concrete should not be used for loads in excess of those which might safely be applied to the concrete of the thickness or form used, without reference to the steel introduced therein. Not only is the expanded metal too thin for permanent construction, but, furthermore, it is so cut into shreds as to expose both surfaces and edges to corrosive action, and this within a material, cinder concrete, about which serious doubts exist as to its corrosive tendencies. Many examinations have been made during the past few months of expanded metal which has been embedded in concrete since this system was first introduced six years ago, and although no reports have been published of seriously corroded material, allowances must be made for the relatively short trial which the system has had. Five or six years cannot be considered as an adequate test for permanent construction.

The arched system overcomes these objections, but is weak in the usual lack of protection given to the lower flanges of the beams.

Columbian Floor.—It is to be regretted that the New York tests did not include a sample of the Columbian flat-ceiling construction. The floor tested was of the panelled construction, without ceiling protection of any kind, and although the results showed stone concrete to resist successfully a long-continued fire, the test was not a satisfactory demonstration as warranting the use of this system. The concrete sank in places, and many of the bars were exposed after the water-test. The slab-casings under the beam-flanges were destroyed, and sufficient heat was transmitted to the surface of the floor to cause the wood screeds to smoke.

Had the flat-ceiling construction been employed, the result would doubtless have been far better.

Roebling Arch.—The fire- and water-tests on this floor are very satisfactory. The weak points have been shown to be the ceiling construction, and the absence of protection to the lower flanges of the I-beams. The arch formed permits of a ready reconstruction, and tests have demonstrated the strength to be all that could be desired.

Aside from the question of cost, commendable features or qualities are possessed by the Roebling and Expanded Metal arched systems, but for first-class fireproof construction neither the Roebling nor the Expanded Metal arch is to be recommended without the adequate protection of the lower flanges of the supporting beams, *in addition to* the suspended ceiling. All those parts of the beams which project below the main arch should be fireproofed just as thoroughly as though the suspended ceiling were not used. Suspended ceilings will retard the heating of the load-carrying arch and steel beams, and will cause the fireproofing above to be less liable to serious damage by fire and water, but the protection thus afforded is not alone sufficient for absolute reliance.

Concrete versus Terra-cotta Floors.—As to a choice between concrete and terra-cotta floors, personal opinions will be found to vary widely. Thus in the two reports on the Pittsburg fire, made by the appraisers and by the board of engineers, two distinct and conflicting recommendations are made regarding concrete and terra-cotta. In the appraisers' report it is stated that " We believe that this same class of material (hard tile) of the end-construction made heavier throughout, with an extra web to strengthen it, and properly put up, would stand any ordinary kind of fire. The absence of panels in the ceiling, we think, would also be an advantage, as the plain ceiling stood all right."

The engineers' report contained the following: " The most important lesson taught by this fire was the lack of strength developed by the fire-clay fireproofing. The building was permitted to move in any direction without any restriction by the fireproofing. The floor-arches showed by the scaling off of the lower webs that they were unable to offer any sufficient force to counteract the tendency to lateral motion.

" In view of these important developments, it is our opinion that important structures of this class should have a radically different method of fireproofing. The fireproofing should be in itself strong and able to resist severe shocks, and should, if possible, be able to prevent the expansion of the steelwork.

" There seems to be but one material that is now known that could be utilized to accomplish these results, and that is first-class concrete. The fire-resisting qualities of properly made concrete have been amply proven to be equal to if not better than fire-clay tile, as shown by the series of tests carried on by the Building Department of the City of New York.

." From the experience gained in street-railway construction in laying continuous rails, it is to a large degree possible to prevent the metal from expanding. In street-railway work this has been accomplished merely by the adhesion of the pavement to the side of the rails. In building construction the same results could be obtained by encasing the columns and girders with concrete placed directly against the steelwork. The adhesion of the concrete would to a large degree prevent unequal expansion of the concrete and steel. The floor arches should also be constructed of concrete, but of sufficient depth to be able to resist lateral forces. With the prevention of injurious expansion and the protection of columns with materials that can stand severe shocks of any nature whatever, the modern steel-frame constructed building would be more thoroughly protected against fire.

" The construction herein suggested should not materially increase the cost of construction. The solid concrete about the columns would be added strength to same and could, no doubt, be made self-supporting. The same could be said of concrete surrounding beams and girders, as has been amply demonstrated by the strength developed by concrete-iron constructions such as the Monier and Melan arches.

" One of the objections that would be raised against this construction would be that it could not be carried on in freezing weather. This is a proper objection, but there is no valid reason in our opinion why these structures should be built in winter any more than were the old-time brick structures."

These recommendations do not point to either one or the other system as the *only* satisfactory construction. If the tiles are made of thick porous material, as recommended in Chapter VIII, with rounded interior corners to prevent cracking, deep skew-backs, and arches with level ceilings, and concrete filling of *good quality*, terra-cotta floors would accomplish all the ends called for in the criticisms of the board of engineers.

The tests made by the New York Building Department and the action of the floor-arches in the Home Insurance Building fire certainly do not show in the least that terra-cotta is an unsuitable material for fireproof construction, while, on the other hand, the New York tests should satisfy those who have heretofore discriminated against concrete construction.

In the volume entitled "Concrete" by Thos. Potter, attention is called to the fact that the ordinary strength of good concrete is so great that if it will carry a specified live-load at the end of a month, at the end of a year it will probably support three times as much. Hence the loss of strength under heat, as indicated by experiments, and these for tensile strength only, is not so vital as would appear. Mr. Potter states, as his conclusions (page 180), that if it is more convenient or otherwise desirable to use concrete, this material may be used for floors and it will be fire-resisting to all intents and purposes for domestic and public buildings containing no great amount of combustible materials, or for almost any structure, except the warehouse class, which contains large undivided areas, or large stocks with heavy loads. When used, the concrete floors should be of the best materials and workmanship, completely encasing the iron- or steel-work, and in the shape of an arch, capable of carrying at least three times the live- and dead-loads the floors were designed to carry.

Commenting upon the New York tests, the *Engineering News,* in an editorial dated Feb. 4, 1897, said:

" Our knowledge of both the strength and the fire-resisting qualities of tile floors is confined almost wholly to the results of experimental tests. Many of these have been made, and their results may be fairly summed up by saying that they have shown hollow tile arches of good design and not too long span to have a strength sufficient to be safe under any load likely to be brought upon them in any ordinary building; also, that their resistance to fire is entirely satisfactory, except that the dense tile is likely to go to pieces if struck by a stream of water when heated. This has been shown by actual fires in fireproof buildings of tile construction, as well as in experimental tests.

" Turning now to floors made up of cement concrete and

steel, we know that the extensive use of this material in Europe has nearly as long a record as has the use of tile here. We know that properly designed arches of this material have an enormous strength, much above that of tile arches of equal weight and span.   Until recently, however, there has been a rather small amount of knowledge as to the fire-resisting qualities of such constructions, and while many able engineers asserted that concrete would resist fire, other authorities denied it.   We may fairly say, however, that this question has been effectively settled by the series of tests inaugurated by Supt. Constable of the New York Building Department, and the engineering profession and the building trades owe him a debt of gratitude for this work.   Some half dozen tests have now been made of concrete floor constructions, and, although the conditions of the tests are exceedingly severe, not one of these floors has failed.''

The general conclusion is therefore warranted that either construction, within the limits pointed out, is good if well executed, while both are bad where defective materials or workmanship are employed.

CHAPTER X.

COLUMN FIREPROOFING.

Importance.—The most important load-bearing members in modern steel-building methods are, without question, the metal columns. Interior columns, which are here especially referred to, stand isolated and exposed on all sides, but form the supporting members for areas which, when fully or even partly loaded, induce strains of remarkable degree; and in buildings of great height, or of very heavy loading, the summation of the loads in a column-shaft often produces strains which but relatively few years ago would have been considered visionary. In general it may be said that the higher the building the greater become the column loads; but in special construction, through the introduction of heavy plate- or box-girders, or through the use of trusses to carry floor- and column-loads over a clear space beneath, the heavy concentrated loads resulting from such construction may exceed the loads found in even the highest buildings. In the Park Row Building in New York City, thirty stories in height, the heaviest column load is 2,900,000 lbs., while in the Waldorf-Astoria Hotel, a column supporting the large trusses over the ballroom carries a load of 5,400,000 lbs.

The importance of properly fireproofing a column or structural member increases in proportion to the service rendered. Practice shows that this point is very often overlooked. The steel-frame is carefully designed for the required dead- and live-loads, and the individual members are accurately proportioned for recognized fibre-strains computed by accepted formulas; but from this point on, the proper fireproofing of such members as isolated columns, which demand architectural treatment in many cases, or which must utilize as little floor-space as possible, resolves itself into a question of "how small," rather than "how good." The contractor for the fireproofing is given the dimensions of the finished column, from which must be deducted ¾ in. on all sides for plaster. This often leaves less than 2 ins. for the fireproof blocks, which must even be cut or gouged to keep them within the given architectural dimensions.

The ingenuity and thought expended upon new types of floor-construction, all of which at least aim to protect the floor-beams, is not paralleled by any great improvements in the question of column fireproofing. Many of the companies which furnish and erect patent floor-constructions also have their own system of column fireproofing, but the attention of the architect is principally engaged by the merits of the floor, and the accompanying column protection is often accepted with the type of floor selected. A construction company may control a most commendable type of fireproof floor, while the system of column fireproofing employed by the same company may be poor in the extreme. There can be no excuse for linking one with the other.

No part of a steel building requires more attention as to fireproofing than the columns, and absolutely no considerations of appearance or question of floor-space occupied should be allowed to influence unduly the shape or size of the fireproofing material.

Essentials for Column Fireproofing.—The requisites for acceptable column fireproofing are:

1. Indestructibility by fire or water, or the combined action of both.

2. Permanency of position, so that the covering cannot be dislodged by fire or water.

3. Non-conductivity of heat.

4. Invulnerability at the joints.

The fireproofing should in no way be dependent upon any combustible material. It should be continuous from the floor-plate or floor-arch to the ceiling, resting firmly and directly on the fireproof floor, and not on wooden flooring or on wooden sleepers, as is often done. The material must be adapted to resist both fire and water, or alternate attacks from each, and it must be so securely held in place as to withstand such attacks without dislodgment. If the capacity for long resistance to fire is to be developed, or if the member is an important one in the structural design, two thicknesses of any material placed in block form should be used, the different layers breaking joints. The material employed must also be non-heat-conducting, so as to preserve the metal work from undesirable expansion.

Metal Lath and Plaster Protection. — Stiffened wire-netting, Bostwick lath, and expanded metal lath are exten-

FIG. 77.   FIG. 78.
Lath and Plaster Column Protection.

sively used with plaster coatings as a means of column protection.

In many instances the column, especially if circular in form, is simply wrapped close with metal lath, and plaster is then applied without any intervening air-space. A better method is to use some form of furring strips to separate the lath and plaster from the column, as shown in Fig. 77. The furring strips are often made of small V-shaped pieces of sheet-iron, placed in a vertical position around the column, to which the wire-netting or lath is wired.

Stiffened wire-netting or metal lathing is sometimes wrapped around and fastened to a light framework or furring of rods, light angles, or channels, as indicated in Fig. 78. The furring is held clear from the column by means of straps or clamps.

The best method of column protection by means of plaster is through the use of a double wrapping, with an intervening air-space. (See Fig. 79.) Metal lath is first wrapped around the column and securely wired at the lap. After a heavy coat of hard mortar has been applied to this wrapping, $\frac{3}{4}$-in. V-shaped furring strips are placed, upon which is wired a second covering of metal lath, which receives two rough coats and one finished coat of plaster.

FIG. 79.—Lath and Plaster Column Protection.

Metal lath and plaster protection may suffice in moderately severe fires of comparatively short duration, but the employment of these details can be considered only as cheaper substitutes for better methods. The fire in the Livingston Building in New York (see Chapter III) showed that portions of metal lath and plaster protection may stand for some time under severe conditions, while other portions are completely destroyed, leaving the columns exposed. The method illustrated in Fig. 77 is particularly to be avoided, as the close wrapping of the metal lath does not leave sufficient room for a key to the plaster on the under side. The plaster will soon fall off under the action of fire or water. The detail shown in Fig. 79 is much to be preferred to any cheaper form of metal lath and plaster protection, but this will cost little, if any, less than a still more efficient covering of concrete or terra-cotta. Fire-tests of suspended ceilings and metal lath and plaster column protections and partitions have shown pretty conclusively that these constructions will not stand severe applications of fire and water. Their use is therefore not to be recommended.

Concrete Column Casings.—A number of the various companies that manufacture concrete floors employ solid concrete as the means of column protection.

Columbian Fireproofing Co.'s Method.—For open forms of columns, such as the Z and channel shape, etc., wooden nailing-strips are first wedged in between the flanges at about every two feet centres. To these are nailed long sheets of tin with locked joints, the purpose being to prevent the concrete from filling the spaces between the flanges of the open column section. Wooden boxes are then made at the building, of the required finished form of the concrete casing, the boxes being made in two sections or halves, each of the full length of the column. The concrete is lifted to a staging, and is poured into the boxes from the top. The column-casings are always finished before the floor above is set, and they are placed directly on the concrete floor below. See Fig. 80.

The concrete employed is made of 1 part Portland cement, 2 parts sand, and 4 parts small broken stone.

The brackets on the columns, for the girder-seats, are also thoroughly encased in concrete, giving the effect of corbels.

Roebling Method.—In this system all free columns are first encased with painted wire lathing, stiffened with $\frac{1}{4}$-in. solid steel ribs or rods woven in every $7\frac{1}{2}$ ins. Suitable light iron furrings are placed to receive the lathing, so as to offset the lath at least 2 ins. from the metal column. This furring may be made circular, square, or any special shape desired. The space between the column and the wire lathing is then filled in solid with concrete, and the surface is given a coat of finished plaster. See Fig. 81.

FIG. 80.—Columbian System of Column Protection.

FIG. 81.—Roebling Concrete Column Protection.

In the Druecker warehouses, Chicago, built in 1898, the columns were fireproofed as follows.* A concrete composed of 1 part cement, 1 part lime putty, and 4 parts cinders was well rammed into wooden forms placed around the columns. These forms were cylindrical in shape, made of 2-in. staves, and in sections 4 ft. long. They were hinged to open in the direction of their length. (See Fig. 82.) The concrete was poured in place from the top, and was well rammed, so as to fill completely the inner cavities of the columns, as well as to surround them entirely. As soon as one 4-ft. section was concreted, a second section was constructed on top of the first, and this process was continued to the top of the column, before the floor was placed. This insured a continuous envelope to the column, without joint from basement to roof. The metal columns were not painted, but were simply cleaned of mill-scale and other foreign substances at the building before the concreting was started. After the

---

* See "Fireproofing of Warehouses," by Frank B. Abbott, *Journal of the West. Soc. of Engrs.*, April, 1898.

floors were laid, the concrete column protections were covered with metal lath, on which was placed a thick coat of dense hard mortar. The metal lath was used to provide a better key for the mortar finish.

Fig. 83 shows the detail of column protection employed in the U. S. Appraisers' Warehouse, New York City. The

FIG. 82.   FIG. 83.

FIG. 82.—Concrete Column Protection and Circular Forms.
FIG. 83.—Column Protection in the U. S. Appraisers' Warehouse, New York.

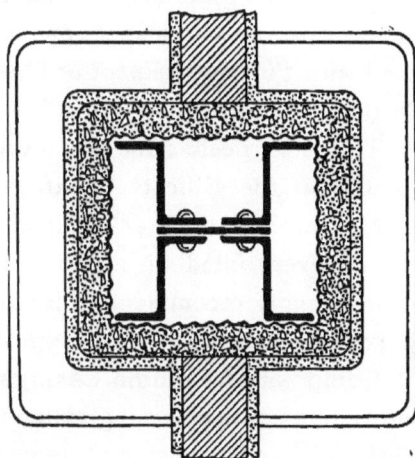

columns, which are cast-iron in the basement and lower stories and Z form in the upper stories, are surrounded by an envelope of No. 24 expanded metal, $2\frac{1}{2}$-in. mesh, enclosing an air-jacket. This envelope received on its outer surface a 2-in. layer of fine concrete made of 1 part American Portland cement, 2 parts sand, and 4 parts $\frac{1}{2}$-in. broken stone, the outer surface of which was finished with a $\frac{1}{2}$-in. protective coat of asbestic plaster.

Protection of Column Interiors.—In Chapter VI the advisability of using some form of open column was pointed out, the idea being to avoid interior surfaces which cannot be readily inspected or protected against corrosive influences.

If box cr closed sections are used, it was suggested that the interiors be filled with some protective material.

Mr. J. Newman, in his " Metallic Structures," states that the closed or cellular form of member is not now often used in Great Britain, due to the difficulty of repairing and to the impossibility of interior painting. He advises that, if a box-girder, column, or any cellular section has to be adopted, thereby preventing inspection and internal painting, it be lined with Portland cement or filled with a comparatively rich Portland cement concrete.

This has been done in several cases in this country, notably in the Ellicott Square Building, Buffalo, N. Y.— D. H. Burnham & Co., Architects—where all of the exterior columns were filled with Portland cement concrete. This is to be strongly recommended for all closed columns occurring in exterior walls or in other exposed locations.

Composition Column Casings.—There are several different systems of fireproofing columns with composition slabs of plaster of Paris, strawboard, etc., a typical example of which is given by the Mackolite type.

For round columns, semi-cylindrical sections are used,

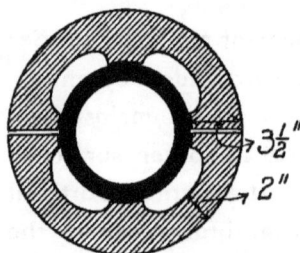

FIG. 84.—Mackolite Column Protection.

having inside projections which bear against the column, forming hollow spaces. See Fig. 84. The sections are held together by steel clips. For rectangular casings, partition blocks are used, built around the member to give the required form.

The conductivity of heat of this material is very low, but it will not resist abrasion nor the disintegrating effects of water.

Terra-cotta Column Casings.—*Cast-iron Columns, Circular Finish.*—Architectural effect or the amount of space occupied by columns often tempts architects to call for very thin solid slabs, so that, for instance, an 8-in. circular cast-iron column may finish, when plastered, not over 12 ins. diameter. This requires blocks of about 1-in. thickness, and the use of such tiles should never be permitted. They are too thin to form even passably good protection, and it is impracticable to manufacture them successfully. They must be dried over a drum, and in doing this the shrinkage causes them to crack through the centre, unless the pieces are made very small, in which case the setting becomes impracticable.

The most common form used for circular columns is shown in Fig. 85. This is made up of segments, struck on the required radius, having protruding ribs on the inside which fit close against the column, thus forming voids when placed in position.

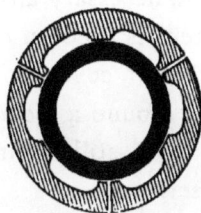 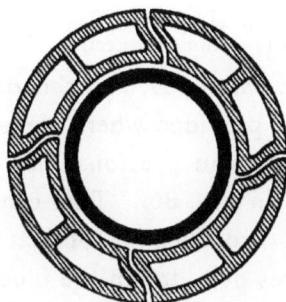

FIG. 85.   FIG. 86.
Terra-cotta Column Protection—Circular.

Fig. 86 shows a casing made of hollow blocks, with lap joints.

Circular column casings are laid in layers or courses, each course being held in position by a binding with copper or soft steel wire. The upright joints are broken.

It has been suggested to make blocks with dovetailed vertical joints, as shown in Fig. 87. The dovetails would run up and down, for the entire width of the blocks. The courses could be locked one to another by rabbeted joints along the horizontal edges.

Where rectangular cast-iron columns are to be encased in circular form, the same blocks are employed as used for circular columns. The diameter of the circular covering is made sufficiently large to fit the longest diagonal of the column. The form shown in Fig. 88 is often used, but it is not suitable. To make the casing rigid, a solid backing of mortar

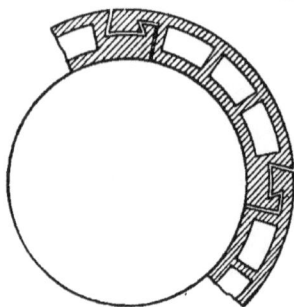

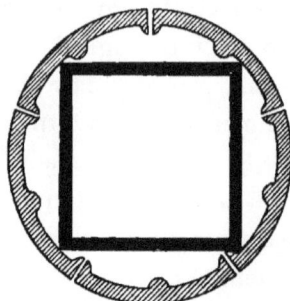

FIG. 87.—Dovetailed Terra-cotta Column Protection.   FIG. 88.—Terra-cotta Column Protection—Circular.

must be employed, and this destroys the necessary air-space, which is provided where the same system is used on circular columns, as in previous Fig. 85. A better covering would be that in Fig. 86. This can be fitted around a rectangular column, and filled in solid at the back, but still leaving the air-spaces provided in the blocks themselves.

Steel Columns, Circular Finish.—The most common form is as shown in Fig. 89, but unless the column is shaped out by mortar, concrete, or additional terra-cotta blocks, so as to give a circular form before the casing is placed, the previous remarks about rectangular cast-iron columns with circular coverings are applicable. To be effective and rigid, the blocks must either have a solid filling or else they must bear against the column.

The best circular form for a Z-bar column is illustrated in Fig. 90, where the terra-cotta blocks are made to fit all standard sizes of columns. The diameter of the casing is obtained by adding $1\frac{1}{2}$ inches to the greatest diagonal of the steel column. This will give the smallest possible effective covering, but if space will permit, an increase in the diameter will give a better protection.

In setting this covering it is impossible to avoid continuous horizontal and vertical joints, but the shape of the casing and its method of application insure sufficient strength and

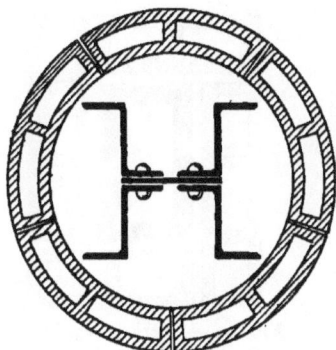

FIG. 89. FIG. 90.
Terra-cotta Column Protection—Circular.

rigidity without additional fastening, provided the blocks are well bedded in cement-mortar. But to guard against bad workmanship, as frequently occurs, it is well to make the invariable rule that all casings should be bound or wrapped with wire, at least one wire to each horizontal course.

If it is desired to improve on this design, a second casing 2 ins. thick could be added around the exterior, leaving a 1-in. air-space between this and the inner protection.

A circular casing for a channel column is shown in Fig. 91. This detail was employed in the New York Life Building, Chicago, Jenney & Mundie, Architects. Each tile was clamped to those above and below, as well as around the column. The blocks were set in mortar made of 1 part Portland cement to 3 parts best lime-mortar.

Cast-iron Columns, Rectangular Finish.—The most common method is to use 3-in. partition blocks, laid up to give the required outside dimensions. (See Fig. 92.) Various

thicknesses of blocks are used, but 3 ins. is the usual practice.   The blocks are set so that the vertical joints alternate
in the successive courses.   This is usually considered to give
sufficient bond to hold the blocks in position, and if the work
is done carefully it should be satisfactory.

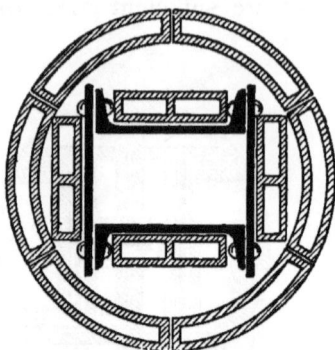

FIG. 91.—Terra-cotta Column   FIG. 92.—Terra-cotta Column
      Protection—Circular.   Protection—Rectangular.

A more efficient casing would be obtained by first covering the column with metal 'lathing and a coat of plaster, outside of which should be placed the terra-cotta casing, with a
1-in. air-space between the two.

Rectangular columns are often finished with quartered or
rounded corners, as shown in Fig. 93.

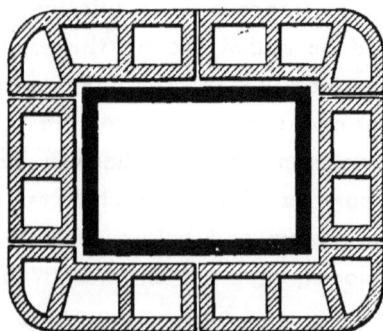

FIG. 93.—Terra-cotta Column Protection—Rectangular.

The importance of fireproofing cast-iron columns is
usually not as great as is the case with steel columns.   Cast-

iron columns will better resist the action of high temperatures, but they are not ordinarily employed in the heaviest
or most important work.

Steel Columns, Rectangular Finish.—The more common
forms are as shown in Figs. 94 and 95, which illustrate square

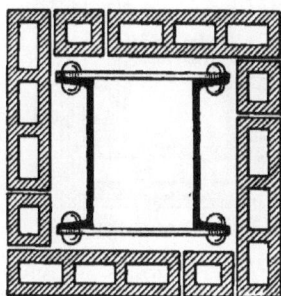

FIG. 94.   FIG. 95.

Terra-cotta Column Protection—Rectangular.

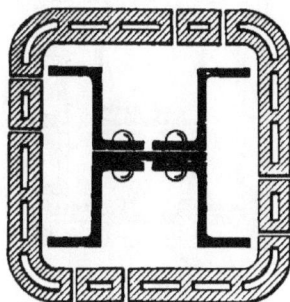

and quartered corners respectively. The blocks employed
are from 2 ins. to 4 ins. in thickness, and they are usually
placed without backing. Old brick or other refuse material
is often filled in behind the casing to block it out and steady it
while being set. Careless blocking in this manner should not
be permitted, but an arbitrary rule should be .established
requiring a careful and solid filling behind all column casings.
This is necessary, not only to preserve rigidity of position,
but to prevent the possibility of draughts or flame working
up any vertical flues within the casings. Another form of
Z-bar column casing is shown in Fig. 96.

A most commendable double casing is illustrated in Fig.
97. This detail was used in " The Fair " building, Chicago,
Jenney & Mundie, Architects. In case the outer layer is
damaged or displaced, the column still has the protection of
the inner blocks. A similar detail was also used for circular
columns, in which case the inner layer was bound in place by
either wires or wire-netting.

An improvement might be effected by keeping the outer 2-in. casing 1 in. away from the inner layer.   Rigidity could

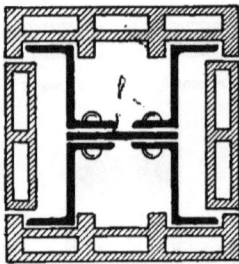
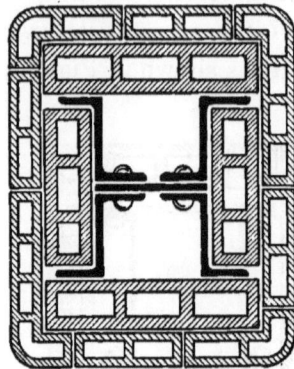

FIG. 96.   FIG. 97.
Terra-cotta Column Protection—Rectangular.

then be provided by using 3-in. blocks in places in the outer covering, to project back through the air-space, and against the inner casing.

Special Shapes of Column Coverings.—From considerations

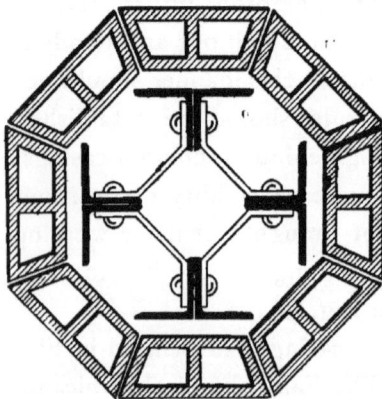
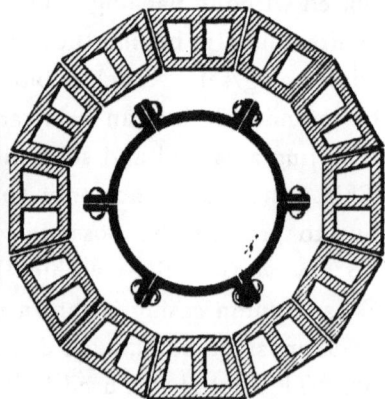

FIG. 98.—Terra-cotta Column Pro-   FIG. 99.—Terra-cotta Column
tection—Octagonal.   Protection—Special Form.

of architectural treatment, special shaped casings are often desired, as indicated in Figs. 98 and 99. Casings of unusual form are generally worked out by the manufacturers in the most practical way, but as these have not received any special study, and as the opportunity of securing knowledge from experience is lacking, their practical value is greatly lessened. For the best work, it is advisable to adhere to the most simple and reliable forms.

Hard versus Porous Terra-cotta Casings.—The constant abrasion and the continual possibility of jars or blows require that the column casings should be made of a material which will withstand these damaging elements, as well as fire. This applies in greater force to columns than to any other portion of the terra-cotta work.

Hard-burned terra-cotta is not a suitable material to use under these conditions. It will not satisfactorily resist shocks, and at time of fire the application of water upon the highly heated blocks will cause them to crack and fall from position. A fire of severe intensity and long duration is very liable to cause the total destruction of a hard-tile casing.

Porous terra-cotta is, from its nature, a poorer conductor of heat, and it undergoes no perceptible expansion under the influence of high temperatures. Its greater elasticity and ability to withstand greater heat enable it better to resist the application of cold water if poured on the casing while hot. Considerations as to the permeability of moisture mentioned in Chapter VI also show that porous terra-cotta is much the better material.

Solid versus Hollow Column Casings.—To insure the most perfect protection to the column, the form should be built out solid to some regular outline with concrete, terra-cotta, or other acceptable material. This is to secure protection against dampness or other corrosive influences, to avoid the possibility of having vertical flues within the casing, and to prevent deformation by the pushing in of the finished covering. This form of filling should then be treated with some waterproof coating, after which the outside casing may be applied, leaving an air-space, at least 1 in. thick between the two. The outer casing, preferably 2 ins. or 3 ins. thick, may be held in position by inserting, at intervals, blocks which will extend in and abut against the core.

In the fire in the Home Life Insurance Building the cement bond of the joints in the terra-cotta column casings was broken in many instances and a resetting of much of the tile was necessitated. A solid backing would have given increased stability to the blocks and prevented the fire from entering the joints.

Pipe-spaces.—As was pointed out in Chapter VII, the proper installation of the mechanical plant becomes of vital importance in fireproof construction. All piping, etc., should be carried in chases or compartments especially designed as pipe-receptacles, these to be always accessible for repairs or changes.

As a matter of economy, both in original cost and in the

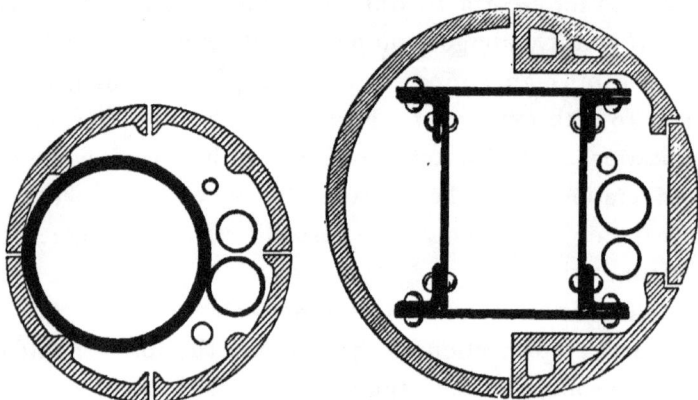

FIG. 100.   FIG. 101.
Circular Column Casings with Pipe Spaces.

matter of space, it has been the common practice to run water-, waste-, and vent-pipes immediately alongside the steel columns, and inside of the fireproofing enclosure, as shown in Figs. 100 and 101. Indeed this practice is specifically advised in several of the handbooks issued by steel companies. Piping is seldom, if ever, placed *inside* of the *steel column*, but a detail almost as bad was used in a case where the horizontal cap-plates of the steel columns were slotted at each floor or joint to allow the passage of the pipes inside of the column area. Open forms of columns, such as the Z-bar and Gray types, have been extensively recommended as giving considerable pipe-space within the minimum circular or rectangular enclosure of fireproofing, but the best practice now condemns the running of supply-, vent-, or waste-pipes within the same enclosed space with the steel column.

Wherever pipes run alongside of the steel columns, they should be separated from the columns by an adequate wall or protection of fireproofing.

The specifications relating to the column fireproofing in the Reliance Building, Chicago—D. H. Burnham & Co., Architects—were as follows:

"All columns throughout the building, including the attic, shall be fireproofed with 3-inch tile, special tiles being used having rounded corners, and provided with slots to receive the pipes where indicated, set plumb to a line, regularly bonded, having air-spaces between the fireproofing and the iron, keyed in place and to each other, and each piece wired to the column with copper wire. Wherever pipes are run alongside of columns, they are to be separated from the column by fireproofing." * (See Fig. 125).

In the newer portion of the Monadnock Building, Chicago, the columns were fireproofed as shown in Fig. 102. " Hollow bricks, laid in cement-mortar, were built solidly around the columns to a line distant 4 ins. from the extreme points of the metal work, and a 2-in. coating of hollow tile was then

---

* See "Steel Skeleton Construction," by E. C. Shankland, Min. Proc. Institute C. E., Vol. cxxviii, Part II.

laid against the brick backing, and extending beyond the
column in one direction to serve as a space for the reception
of vertical pipes.'' *

Practically the same result may be obtained by using a
solid casing of porous terra-cotta around the column in place
of the hollow brick.   In cases where independent and accessi-

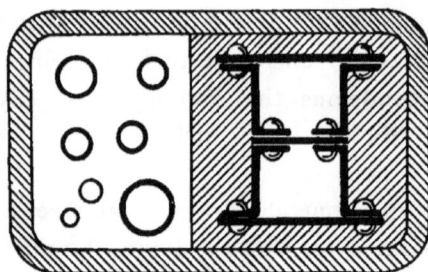

FIG. 102.—Brick Column Casing and Pipe-space—Monadnock Building,
Chicago.

ble pipe chases are not provided, this detail will be found as
satisfactory as any that can be used.   In circular columns,
the outer casing may be made eccentric to provide the necessary room.   This method will protect the column against
attack by fire, and at the same time prevent deterioration
from corrosion due to the immediate presence of piping near
the metal members.   A coating of some waterproof material
over the inner solid casing becomes of especial importance
where piping is run in this manner.

Column Guards.—In mercantile or storage buildings,
where hand-trucks are used in transferring merchandise, the
column shafts should be protected to a height of from 5 to 8
feet by wrought-iron or steel column guards.

These are usually made of $\frac{5}{16}$-in. or $\frac{3}{8}$-in. plates, iron or
steel, spliced vertically.   They may be either fastened to the
finished floor or to the rough underflooring.   The guards

* See '' Architectural Engineering,'' J. K. Freitag, page 134.

are usually made with slightly rounded corners for rectangular casings, and of circular form for circular columns, and of size to fit closely around the fireproofing. The plaster finish is then applied to the fireproofing, coming flush with the metal guards.

A small cast moulding is sometimes attached to the upper edge of the guard to conceal the joint between the guard and the plaster.

These metal casings form a valuable protection against damage from passing trucks, falling packing-cases, etc., all of which tend, in time, to loosen or even break the fireproof covering.

CHAPTER XI.

FIRE-RESISTING PARTITIONS.

PRESENT methods of so-called fireproof partition construction undoubtedly constitute the weakest feature in modern fire-resisting design. A careful inspection of the many systems of fireproof partitions as used in most of our large and pretentious steel buildings, which are supposed to be thoroughly fireproof, will reveal the fact that one of the most important functions of the dividing partition, viz., its fire-retarding qualities, has either been overlooked, or else considered of such small importance that architectural requirements have been allowed to overrule the matter of the service of such partitions in case of conflagration.

The average partition is nothing more than a light screen, erected to subdivide areas according to the demands of tenants, to form corridors, or enclose stair- or elevator-wells, or make other architectural subdivisions. These partitions often contain large areas of interior sash for the transmission of borrowed light to or from the corridors or from one room to another, besides considerable quantities of wood in doors, door frames, and even in wainscoting. Thus in considering the architectural purposes for which the partitions were erected, the equally important question of fire-retarding capacity is largely lost sight of.

Requirements.—The requirements demanded of a thoroughly fireproof partition are:

236

1. Architectural service, in securing a convenient sub-division of areas.

2. Fire-resisting service, as limiting the spread of fire.

3. Heat-retarding qualities, as preventing the passage of heat to adjoining areas.

4. Stability against shock, water streams, etc.

5. Deadening qualities, to prevent the transmission of sound.

The architectural arrangement of partitions, though placed first as regards the requirements, should be considered with, and as a part of their service in preventing the spread of possible fires. In Chapter VII the value of the subdivision of large areas was pointed out, as serving to confine the fire to its original locality, and to permit more ready handling by the fire department. The partitions should therefore be located with this definite object in view, and if adequate brick or other fire partitions are placed at suitable dividing lines, less efficient partitions could be placed within the areas so formed.

Upon the partitions will fall the responsibility of localizing the fire in the compartment in which it originated. If the partition is properly designed, it will confine the damage due to fire or water to the compartment in which the fire started, and cases have even been recorded in which fire broke out at night in efficiently enclosed areas, and its occurrence was not known until the following morning.

In addition to preventing the passage of flames, the partition must be sufficiently non-heat-conducting to insure the protection of adjoining areas from radiated heat, and sufficiently stable to withstand the jars or shocks due to ordinary service or to the force of fire-hose. The importance of deadening qualities will largely depend upon the use for which the partition is intended, but in buildings containing many tenants this is an important consideration.

The more ordinary methods of partition construction will
be described and a comparison of their various good and bad
features will be made.

Solid Plaster Partitions.—A solid plaster partition having a total thickness of 2 ins. is used by the John A.
Roebling's Sons Company. The studs consist of either $\frac{7}{8}$-in.
channels or 1-in. by $\frac{3}{16}$-in. flats, spaced 16 ins. centres.
These are secured at the bottom, by nails or staples, to a
$\frac{1}{8}$-in. by 1-in. wood strip, which is laid on top of the concrete
floor-arch or slab, and at the top by staples driven into the
ceiling.

Wire lathing, stiffened with $\frac{1}{4}$-in. solid steel wires or ribs,
woven in every $7\frac{1}{2}$ ins., is laced to one side of the studs with
No. 18 galvanized wire. The sheets of lathing are so placed
as to make the stiffening ribs run at right angles to the studs.
See Fig. 103.

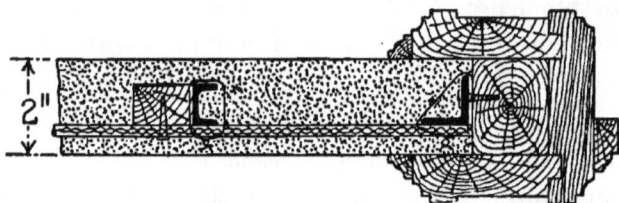

FIG. 103.—Two-inch Solid Plaster Partition.—Roebling System.

All openings for doors, transoms, windows, etc., are
framed with 1-in. by 1-in. by $\frac{3}{16}$-in. angles or by means of
$\frac{7}{8}$-in. channels. The vertical members at door-openings are
made to extend the full height from floor to ceiling. Such
members around openings are punched at intervals with holes
to permit the fastening of the wood frames, etc.

Wood furrings, $\frac{7}{8}$ in. thick, are placed between the studs
to receive the base-board, chair-rail, and picture-moulding.
These furrings are usually placed by the carpenter, and are
held by staples going around the studs or around the $\frac{1}{4}$-in.
rods.

The partition is then plastered with some hard plaster or lime-mortar gauged with plaster of Paris. This is applied in five coats, giving a total thickness to the partition of 2 ins. The weight including plaster is 20 lbs. per square foot.

Another form of solid partition used by the same company is made of a combination of cinder concrete and plaster on an iron framework. The studs are made of 2-in. by $\frac{1}{8}$-in. channels, spaced 18 ins. centres, extending from concrete floor-plate to ceiling. They are fastened by bending the ends of the channel-webs to form small knees, which are spiked to a 2-in. by $\frac{1}{8}$-in. wood strip, nailed to the concrete floor. Openings are framed by means of 2-in. by 2-in. by $\frac{1}{8}$-in. angles, or by channels of the same size as the studs, punched to allow the securing of wood frames. Upright angles at sides of doorways extend the full height of the partition.

After all of the iron framework is in position, with the necessary wood frames, stiffened wire lathing is laced to both sides of the studs, with the $\frac{1}{4}$-in. stiffening ribs, spaced every $7\frac{1}{2}$ ins. centres, running at right angles to the studs. The space between the two surfaces of wire lathing is then filled solid with a cinder concrete composed of 1 part Portland cement, $2\frac{1}{2}$ parts sand, and 6 parts cinders. This completely embeds the studs within a concrete slab about $2\frac{1}{2}$ inches thick. Two coats of plaster, a brown coat and a finishing coat, are then applied to the outer surface of the lathing, making a partition of a total thickness of 4 ins. This construction is shown in Fig. 104. No wood furring is necessary for attaching the base-board, chair-rail, or picture-moulding, as the cinder concrete will receive nails. The weight per square foot, including plaster, is 32 pounds.

For both of these forms the light channels or angles are cut and fitted at the building as required. The wire lathing is furnished in rolls 150 feet long. One of the various hard plasters is usually employed, King's-Windsor, Adamant,

Acme, or Rock Wall, there being apparently no difference in
the fire-resisting properties. Temporary bracing during
plastering is not necessary, although it is sometimes resorted
to, and even required by architects.

A $1\frac{1}{2}$-in. solid plaster partition is made by the Expanded
Metal Companies, consisting of $\frac{5}{8}$-in. or $\frac{3}{4}$-in. upright channel

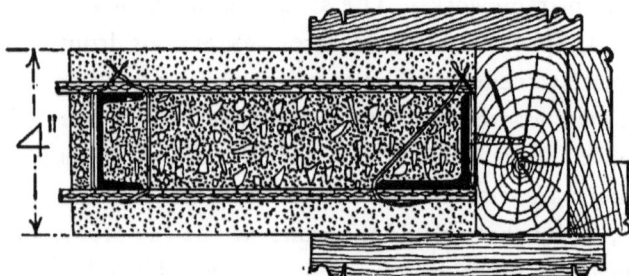

FIG. 104.—Four-inch Solid Plaster Partition.—Roebling System.

studs, to one side of which expanded metal lath is wired.
The studs are placed 12 ins. centres for a height of 10 ft. or
more, and 16 ins. centres for a less height. The ends of the
channel-webs are bent to form small knees, which are secured
to the floor and ceiling by means of nails, screws, toggle-
bolts, or other fastenings, depending upon the conditions
encountered. The sheets of metal lath are wired at least
four times to each stud, and also at the laps, No. 18 or
No. 20 soft wire being used for this purpose.

For a temporary bracing during plastering, $\frac{3}{4}$-in. horizontal channels are wired to the studs on the side opposite the
lathing, one brace being used in low partitions, and two rows
in partitions over 12 ft. in height. The plaster is first
applied to the lath side. After this coat has set, or in about
a day's time, the back side of the partition is plastered out
to the face of the studs. The bracing is then removed by
cutting the wires and the surface is patched where the horizontal channels occurred. Patent or hard plasters are always
used, as common mortar requires too long a time for thorough

drying, and is not sufficiently rigid. In partitions over 12 ft. high the total thickness is usually made 2 ins. The 1½-in. partition weighs about 15 lbs. per square foot.

For these various forms of solid plaster partitions a cement base-board is sometimes moulded in position by the plasterer while applying the finish coat. A cement floor-strip may also be made, extending 12 ins. or 18 ins. out from the partition.

Hollow Plaster Partitions.—A hollow partition, finishing 4 ins. thick, is constructed by the John A. Roebling's Sons Company, as shown in Fig. 105. The studs are made

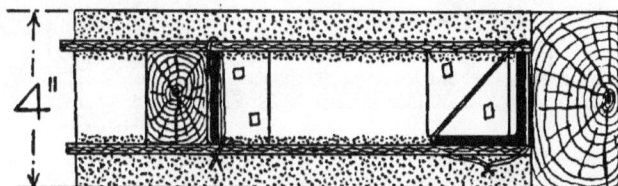

FIG. 105.—Hollow Plaster Partition.—Roebling System.

of 2-in. by ⅛-in. flats, spaced 18 ins. centres, fastened top and bottom by means of knees bent at the ends of the studs. A wooden nailing-strip is run along the partition line on the fireproof floor. Angles or channels are used to frame for all openings.

Stiffened wire lathing is secured to both sides of the studs to receive the hard plaster, which consists of three coats on each side—scratch, brown, and finishing coats. Wood furring-strips are attached to the studs, or to the wire rods woven into the netting, by means of staples.

The weight per square foot is 22 lbs., including plaster.

A type of hollow plaster partition erected by the Expanded Metal Companies is made of 2-in., 3-in., or even 4-in. channel-studs, spaced about 16 ins. centres. These are fastened top and bottom to channels laid along the partition lines, and are stiffened by means of horizontal members or bridging cut in between the studs. Metal lath is then laced to both sides of the uprights, which, when plastered, leaves an air-space over the entire area.

Another hollow partition constructed by the same companies is made of studs which consist of two bars of light angle-irons, riveted together with pieces of light strap-iron every 2 ft. or 3 ft. in height. (See Fig. 106.) For ordinary

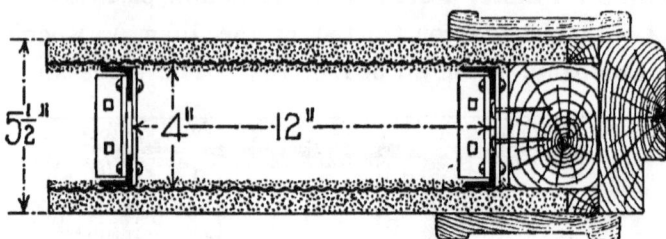

FIG. 106.—Hollow Plaster Partition. Expanded Metal Co.'s System.

partitions, the angles are generally placed 4 ins. out to out, which, with the lathing and plaster on both sides, makes the total thickness $5\frac{1}{2}$ ins. The studs are set 12 ins. centres.

The studs are held top and bottom by means of straps or angle-knees, which are bolted to the studs and nailed directly to the concrete floors and ceilings. A $1\frac{1}{2}$-in. slotted hole is left in the top knees, to allow for inequalities in height. Diamond-mesh lath, No. 24 gauge, is usually employed, with either hard plaster or common mortar.

This detail may be used for any thickness of partition, to enclose vent-flues, heating-pipes, or other features which may require a thick double partition.

Value of Plaster Partitions.—In some of the plaster partitions before described it is to be noted that the iron studs rest upon a wood strip fastened to the floor. The wire lath and plaster extend from the ceiling to the fireproof floor below·the wood strip, which is supposed " to give a bearing for the plaster on the fireproof floor, and to provide a cushion for the studs, so that they can expand in case of a hot fire."

Also, in all partitions except the one which is filled with cinder-concrete, nailing-strips of wood are necessary for the attachment of the base, chair-rail, picture-moulding, etc.

In the light of past experience, these details cannot be regarded as thoroughly fireproof, and for a building intended to possess permanent fire-resisting qualities these constructions are rather to be condemned. The behavior of the partitions in the Methodist Building in the Pittsburg fire showed quite conclusively that plaster on metal lath will not prevent the woodwork behind the plaster from charring, even though the plaster retains its position on the lath. Indeed, sufficient evidence shows that the wood backing will not only char, but that it will burn under a fire of any great intensity or long duration.

This means that the wood strip placed along the floor would act as a "cushion," due not only to its elasticity, but to the charring or burning out under any severe test. This would sufficiently deform the partition to make it unstable and useless in resisting further attacks from fire-hose. The wood nailing-strips would also soon char, and in so doing at least necessitate a reconstruction for the attachment of a new base or other member.

But independent of the employment of wood in such partitions, a sufficient number of fires and tests have been recorded which show conclusively that plaster construction cannot be considered first-class fireproofing. In Chapter V, under a discussion of Mortars and Plasters, the metal lath and cement-mortar partitions in the Home Life Insurance Building were cited as an example. These failed utterly either to confine or resist fire. Indeed, where plaster has occurred in connection with fireproof construction, whether brick, tile, wire lath, or expanded metal, it appears that sufficient bond has not existed between the plaster and the surface to which it was applied to resist successfully the combined action of

fire and water. Lime-mortar in itself is a most desirable
fire-resister, but the action of water, combined with intense
heat, will generally cause it to lose its key and fall from
position, even when sufficient space is left behind the metal
lath to secure the best possible bond.

As regards deadening qualities, solid plaster partitions
are usually very objectionable.

Terra-cotta Partitions.—*Sizes of Blocks.*—The blocks
employed are either square or brick-shaped, according to local
practice or the ideas of the manufacturer.   Square blocks are
commonly made 12 ins. by 12 ins. for the body of the partition, with 6-in. by 12-in. and 8-in. by 12-in. blocks for filling
in the end-spaces, or the tops of the partitions.   For brick-
shaped blocks, a great variety of sizes are used by different
manufacturers, 6-in. by 12-in., 7-in. by 12-in., and 8-in. by
12-in. constituting the more ordinary face dimensions.
Typical partition-blocks are shown in Fig. 107.

Thickness of Blocks.—Partition-blocks are made in thicknesses varying from 3 ins. to 12 ins., 3-in., 4-in., and 6-in.
blocks being the most common.   A 4-in. partition is the most
popular thickness for ordinary work.   With plaster on both
sides, this will finish about $5\frac{1}{2}$ ins. total thickness.   Terra-
cotta partitions of a 2-in. thickness have been placed on the
market, but have not been extensively used.   Their adoption
is certainly due more to faith, or to the claims of the manufacturers, than to any demonstrated qualities of strength.
A 2-in. terra-cotta partition, of any strength or efficiency, is
quite impracticable, and where floor area is so valuable that
more space cannot be occupied, terra-cotta is not the material
to be employed.   Fig. 108 illustrates one type of 2-in. partition in which long strips of band-iron are run horizontally in
the joints to increase the stiffness.   This idea may be plausi-
ble for very short runs, but in considerable lengths, and
especially where the height is equal to or greater than the

length, vertical stiffening members are quite as important as horizontal stiffeners.

Blocks made of porous terra-cotta usually have but one

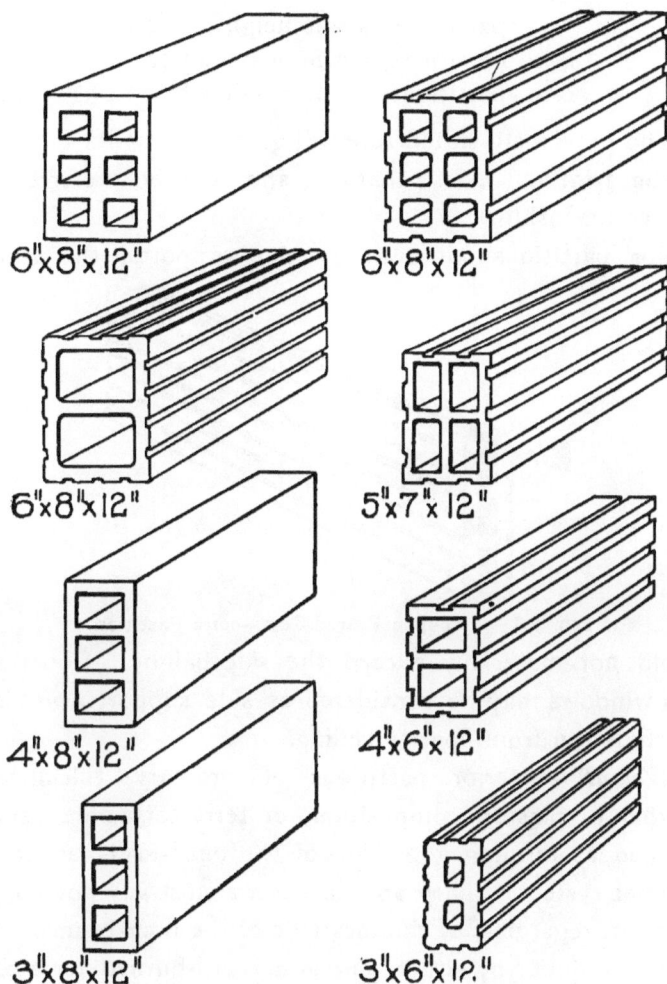

6"x8"x12" 6"x8"x12"

6"x8"x12" 5"x7"x12"

4"x8"x12" 4"x6"x12"

3"x8"x12" 3"x6"x12"

FIG. 107.—Types of Terra-cotta Partition-blocks.

void in the thickness for 3-in. and 4-in. stock. For blocks 6 ins. and thicker, two voids are commonly made. When of the hard-burned material, two voids may be made in blocks of from 5 ins. up.

Height and Length.—The safe height of terra-cotta partitions may be approximated by multiplying the thickness in inches by 40. This will give the safe height in inches. Common practice allows:

> 3-in. partitions, a safe height of 12 ft.
> 4-in. partitions, a safe height of 16 ft.
> 6-in. partitions, a safe height of 20 ft.

The previously stated rule will give less heights than this for the 3-in. and 4-in. partitions, and is to be preferred for best workmanship.

For partitions without any side supports, the length

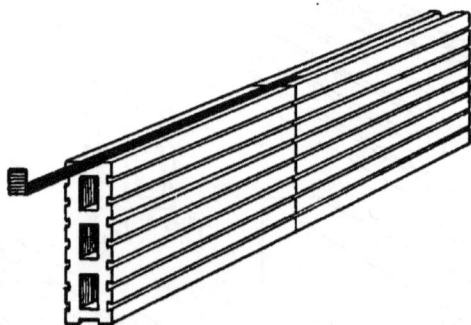

FIG. 108.—Two-inch Braced Terra-cotta Partition.

should not materially exceed the safe height. Doors and high windows may be considered as side supports, provided the studs run from floor to ceiling.

Weight.—Interior partitions of ordinary thicknesses, whether of plaster, composition, or terra-cotta, are usually assumed as forming a portion of the dead-load carried by the floor system. They are placed in almost any position on the floor, regardless of the locations of the floor-beams.

The weights of either porous or hard-burned terra-cotta partitions will average as follows:

> 3-in. partition, 12 to 13 lbs. per sq. ft.
> 4-in. partition, 13 to 16 lbs. per sq. ft.
> 6-in. partition,   22 lbs. per sq. ft.
> 8-in. partition,   28 lbs. per sq. ft.

Method of Setting.—It should be an imperative rule that all partitions must be placed directly upon the concrete, composition, or terra-cotta floors. This will prevent many failures which might otherwise occur.

In the best class of work the lowest course of blocks is either made entirely of *solid* porous material or of solid porous blocks alternating with the ordinary blocks. This is to give a nailing surface for the attachment of the wood base. Above this, the regular partition-blocks are laid, bond broken on all vertical joints, up to the ceiling line. If a chair-rail or picture-moulding is used, these should be provided for by inserting solid porous terra-cotta blocks, to act as nailing surfaces.

At the ceiling, "closures" of the required size are put in, driven as tight as possible, and then made secure by wedging with slate. This wedging is very essential to make the partition rigid and secure against side pressure. Some manufacturers claim that it is best to start setting the partitions in the lower stories first, as the partition weights added to the successive stories above will then cause additional deflections to the beams, and bring added pressure to the tops of the partitions below. Others insist that the partitions should be started at the top first, thus avoiding the increments due to the deflections story by story. If the partitions are well wedged with slate at each story, as should be done in all cases, the best results will probably obtain by working down from the upper floors. Otherwise, sufficient deflection may be obtained in the lower stories to partially crush or buckle the partitions.

If wood studs are used for door or window openings, they should run the full length from floor to ceiling. They should be as thick as the partition-blocks and well straightened, so that grounds may be applied to receive the plastering. Where wood studs extend through the partition, as, for instance, below high window areas, they should be fastened to the adjacent terra-cotta blocks, and then lathed across the face with metal lath of some manufacture which will form a key for the mortar or plaster between the lath and stud. This will allow the stud to shrink without cracking the plaster.

One of the simplest and best ways to secure terra-cotta partitions where they abut against brick walls is to drive large cut-nails into the mortar-joints of the masonry at the top of each course of blocks, before setting the next course. The heads of the nails will then come between the terra-cotta blocks, and by tapping them down with a hammer at the successive courses, great additional stiffness may be obtained.

Rough wooden frames should never be relied upon to sustain the partition-blocks over doors or other openings. If the use of wood frames is insisted upon, however, the partition material should be made to form a flat arch over the openings. Where long rows of corridor lights occur, a metal frame is absolutely necessary to insure any stability whatever in case of fire, and iron frames of some description are much to be preferred for *all* openings, large or small, as will be pointed out later.

Hard-burned terra-cotta blocks should always be grooved or " scored " to provide a key for the plastering. All blocks, whether porous or hard-burned, should be well wet before setting, and again wet before the plastering is applied. Otherwise the absorption will sap the mortar before the cement receives its proper set. The most satisfactory mortar for setting partitions is made of 1 part lime-putty, 2 parts cement, and 2 to 3 parts sand.

Difficulties Encountered in Construction.—The many failures of terra-cotta partitions to withstand fire and water, as demonstrated by fires during recent years in so-called fireproof buildings, have been due to the manner in which the material was used, rather than to the qualities of the material itself. Hardly a single instance can be pointed out in which terra-cotta partitions have undergone a severe fire without suffering very considerable damage, and as this form of material possesses so many admirable qualities, both as regards fire-resistance, strength, and deadening qualities, the necessity for decided improvements in the usual methods of construction is very apparent.

The failures have resulted either from faulty design or from faulty construction.

As regards faulty design, a noticeable example was found in the Home Life Insurance Building. Practically no water was thrown into the upper floors of this building at the time of fire, and yet the heat alone caused the overthrow of many of the hall and room partitions. This was very largely due to the fact that the fireproofing only continued up about 5 ft. from the floor, while above this were placed interior sash to give borrowed light to the corridors and passageways. The following opinion on this feature of design is from a practicing architect: *

"There can be no excuse for the use of such partition-lights in a structure claiming to be fireproof, and the design which requires them for adequate lighting cannot be considered successful."

Many other failures of terra-cotta partitions have resulted from constructive defects, such as partitions left free at the top without proper wedging, or partitions built upon, or made dependent upon wood which has either shrunk sufficiently to cause instability, or burned away in time of fire.

The main difficulties met with in constructing terra-cotta partitions are caused more by the carpenter's work than by the fireproofer's work. Thus for the support of the partitions upon the floor-construction, it has been said that partitions

---

* See W. W. Kent in *The Brickbuilder*, January, 1899.

should always run down to the masonry construction, and not rest upon the wooden flooring. But the carpenter will soon raise decided objections, and use all arguments to avoid this. For if planks fastened directly to the top flanges of the beams are to be used as an underflooring, the carpenter much prefers to lay a continuous floor, and have the partitions built upon this planking. His reasons are:

First, that it is much cheaper for him, as this requires less labor in cutting and fitting, though somewhat more stock.

Second, as many of the partitions, if run down to the masonry, would be placed directly upon the beams or girders, the carpenter would have no fastening for the planking coming against such partitions, but the *next* beam, possibly 4, 5, or 6 ft. away, would have to be used, thus leaving the planks loose at the ends.

If an underflooring of ⅞-in. rough boarding is used, fastened to screeds buried in the concrete filling, the running down of the partitions to the concrete requires the carpenter to place screeds at each side of the partitions, and this is objected to on account of the expense.

If hard-burned terra-cotta only is used in the partitions, a wooden ground is needed for the attachment of the base or wainscot. The carpenter usually prefers to lay a solid nailing-strip the full width of the partition, and have the partition built upon this. Where openings have to be framed for, rough wood frames are provided, but the studs do not extend above or below the doorways or windows on account of the extra expense. Continuous studs at openings add much to the stability during use, but are of no avail during fire. Any woodwork which is relied upon for support is useless under fire.

E ssentials.—Terra-cotta partitions, to be effective under the action of heat and water, should combine the following essentials:

1. All partitions should be started directly upon the masonry at each floor, and be securely fastened by wedging to the ceiling above.

2. If the partitions are to be stable under the action of hose-streams, and the blocks are not of a thickness to secure the necessary strength or rigidity by wedging alone, additional stiffness must be secured by some means or other, or else the thickness of the partition must be increased. A decided step in the right direction to secure additional stability is through the use of tongued and grooved terra-cotta blocks, as shown in Fig. 109. This is an English type, patented by W. Lindsay, and although the tongue and groove may add to the expense of production, the increased strength of the partition would fully compensate for this.

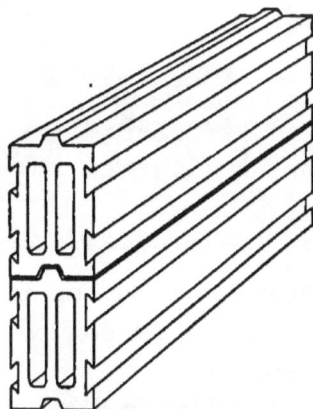

FIG. 109.—Terra-cotta Partition Blocks.—Lindsay Patent.

3. If partitions are intended to be fire-resisting, there can be no excuse for using non-fireproof doors or non-fireproof window areas. To quote again from the motto of a fireproof-door company, " A fireproof door is not intended for a wood partition, neither is a wood door intended for a fireproof partition." The doors should be preferably metal-covered, as described in Chapter VII, and for corridor-lights, if these must be used, the author would suggest the use of wire- or ribbed-glass of fire-resisting ability.

4. If the doors are made fireproof, the door-frames must be made fireproof also, and the same applies to window areas. Fig. 110 shows the arrangement of column casing and partition construction used in the U. S. Public Building at San Francisco, Cal. The partitions are made of 6-in. terra-cotta

blocks with 4-in. cells, all interior door and window openings being framed with 7-in. channels. This detail is to be strongly recommended, as channel-iron frames for the openings will give additional stability, besides forming reliable supports for the terra-cotta work in time of fire. To the channel frames

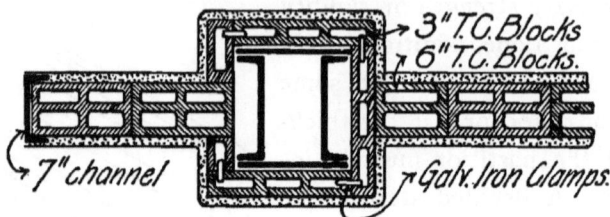

FIG. 110.—Terra-cotta Partitions in U. S. Public Building, San Francisco.

should be attached the metal-covered door or window-frames, for use with the fireproof doors and metal-covered sash and wire-glass.

For thin partitions (either solid plaster or 3-in. terra-cotta blocks) cast-iron door- or window-frames may be used as indicated in Fig. 111.

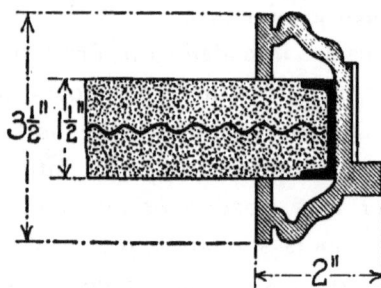

FIG. 111.—Cast-iron Door-frames for Thin Plaster Partitions.

Where absolutely sound-proof partitions are required (as in music studios, etc.), the following specifications have been used by Messrs. Holabird & Roche, Chicago: " Lay up two 3-in. hollow-tile partitions, set absolutely isolated from each other, leaving a 2-in. air-space between. In this air-space shall be hung from ceiling to floor, strips of corrugated building paper, weighing 6 lbs. per square of one hundred feet, perfectly joined at the joints, and continuous from floor to ceiling." By numerous tests, this has been found as nearly sound-proof as any method yet tested.

Composition Partitions. — Mackolite partition-blocks, made of the same materials and in the same manner as described for Mackolite floor-blocks in Chapter IX, are made in much larger slabs than are the ordinary partition-blocks of terra-cotta. The regular size is 12 ins. high by 48 ins. long, the thicknesses being 3, 4, 5, and 6 ins. The blocks are moulded with circular voids extending the full lengths of the blocks, and with grooves along the edges to receive the mortar at the joints. The slabs can be readily cut to any desired shape or width for use as closures, etc.

Increased stability is claimed for this material on account of the large size of the blocks employed, but as the grooved mortar-joints cannot be nearly as strong as the blocks themselves, and as the weight is about 25 per cent. lighter than for terra-cotta blocks, it is difficult to see where Mackolite partitions possess any advantages as to strength. The observations made regarding the essentials demanded of a satisfactory terra-cotta partition will apply equally as well to the Mackolite type. The only advantage which this material possesses is its lightness, and this advantage is questionable when stability is considered. The material will not resist the action of water.

The blocks, unplastered, will weigh as follows:

3-in. partition tile, 10 lbs. per sq. ft.
4-in. partition tile, 12 lbs. per sq. ft.
5-in. partition tile, 15 lbs. per sq. ft.
6-in. partition tile, 18 lbs. per sq. ft.

A plaster of Paris partition, very recently patented by the Metropolitan Fireproofing Co., consists of a combination of

solid partition blocks and metal clips.   The blocks are made
of the usual mixture of plaster of Paris and shavings, as used
in the Metropolitan floor, the size being 2 ft. long, 1 ft. high,
and 2 ins. thick.   The top and bottom edges of the blocks
are rabbeted on each side so as to reduce the thickness of the
blocks to about 1½ ins.   On to these rabbets or grooves are
slipped iron clips, which are stamped out of $\frac{1}{16}$-in. sheet-steel.
(See Fig. 112.)   These clips are placed at the tops and

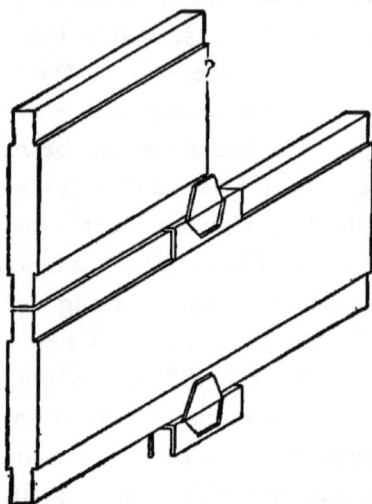

FIG. 112.—Composition Partition Blocks.—Metropolitan System.

bottoms of all vertical joints, and as these joints alternate,
this gives a uniform spacing of 12 ins. for the clips.   The
blocks are made at the factory, and delivered at the building
kiln-dried.   This is to reduce all possible moisture.   With a
plaster finish to each side of the partition, the total thickness
is 3 ins.

CHAPTER XII.

EXTERIOR WALLS, SPANDRELS, AND WALL-COLUMNS.

EXTERIOR walls for steel-frame buildings may be either load-supporting, self-supporting, or dependent for support upon the steel frame.

Before the advent of steel-skeleton construction, the exterior walls were built of solid masonry, and were designed to carry their proper wall-, floor-, and roof-loads, without the aid of metal columns. This still constitutes the ordinary practice in buildings with brick or stone walls of very moderate height, whether of fireproof or non-fireproof construction. Eight or ten stories is the usual height for load-supporting walls, as above this height the piers become heavy, adding materially to the foundation weights, while their bulk consumes too much floor area and reduces the size of the openings required in present practice for ample light and air.

As has been seen in Chapter II, self-supporting exterior walls were employed in the earlier examples of skeleton-construction buildings. The walls served to carry their own weight, and all floor- and roof-loads were supported on metal columns placed within the walls. Self-supporting walls have been employed, in some cases, to a height as great as sixteen stories, and, in buildings of from eight to twelve stories, they have been very extensively used; but the objections of weight and bulk form serious drawbacks, especially in the higher examples of this class.

255

The veneer- or curtain-wall, entirely dependent for support upon the steel frame, has made possible the design and construction of the highest buildings now erected. The masonry wall, which has usually been the most important factor in building construction, now becomes but a veneer, serving as an architectural envelope, and as a protective covering against weather, corrosion, and fire. The walls are supported at each floor-level, thus making a series of walls, each a single story in height. This method results in a very material saving of floor area and weight, due to the reduced thickness of the walls.

From the standpoint of fire-resisting design, the use of the skeleton type of building with veneer-walls is to be preferred, *provided* the walls are made of materials of a character and thickness adequate to protect the steel frame.

In both load-supporting and self-supporting walls, the damage of portions by fire may mean the removal of large areas of uninjured wall at the time of reconstruction, while their load-bearing capacity, especially in the former type, renders them more liable to failure. In load-supporting walls, the bracing necessary to prevent collapse is very important, as failure of the walls would mean great damage to other portions of the structure.

With veneer-walls, parts damaged by fire may be removed and restored with facility, as the method of construction allows the walls to be readily replaced for only such stories or portions of stories as receive injury.

Materials.—The materials used should preferably be brick and terra-cotta, though iron or steel and stone are extensively used. The fire-resisting qualities of these materials have been discussed in Chapter V. In designing and detailing the exterior walls, the following points should be considered:

Ironwork.—Ornamental ironwork is frequently employed,

especially for store-fronts and entrances, where cast-iron pilasters, facias, sills, panels, etc., are required by the design. A very common detail is to make store-fronts run up through two stories, with panelled or ornamented cast-iron pilasters covering the piers, a deep facia at the third floor-level, and panels at the second floor-level between the piers. For such purposes, cast-iron is much to be preferred to steel, as the cast-metal will retain its shape and position far better than thin plates of wrought-iron or steel. All such facings, however, should be so arranged that possible expansion or displacement would not affect the integrity of the structure. An efficient fireproof backing must be placed to protect the structural members behind the facings, in case the latter should fail.

An illustration of the destructive expansion of cast-iron used in exterior work, is found in the action of the cast sills and lintels about the court windows of the Home Life Building. The windows of the light-court were made with cast-iron mullions, sills, and lintels, the horizontal members being about 23 feet long. The heat from the fire was sufficient to expand the sills and lintels enough to force the brick wall, which extended 18 ins. beyond the iron, about an inch out of line, and to develop cracks from this cause.

A very common use of iron in exterior walls is in connection with the light-courts, stairwells, or elevator enclosures, where ornamental cast-iron is employed to form the supporting framework. This usually consists of cast-iron columns, with facias or ornamental panels at the various floor-levels, so arranged as to provide the largest possible window areas, while the entire construction is left exposed to possible external fire. Many examples of such design may be found, in which the sole means of exit, viz., the stairway and elevators, are placed within an unprotected cast-iron framework with large areas of plate glass; and besides, the adjacent buildings may be of combustible construction, and undesirable neighbors even to the best protection. Such practice cannot be considered as fire-resisting design. In case of severe fire in the adjoining buildings, the expansion and warping of the exposed framework and the failure of the window areas is bound to follow, with the ultimate effect of preventing elevator service and causing the distortion of the stairs, if not the complete wrecking of this portion of the structure.

The danger of this form of construction was realized in the case of the Pabst Building, Milwaukee, Wis., where the elevator-shaft was made of an exterior iron framework, exposed to possible attack by fire in an adjoining furniture factory. To prevent the exposed columns and the elevator-guides in this shaft from being damaged or warped by fire, so that the elevator service would be stopped, an 8-in. brick wall was built up as a screen between the elevator-shaft and the wall of the next building. This screen was supported on a plate-girder, and was anchored at intervals to provide stability.

Stone.—Thin slabs of marble, limestone, or granite should never be relied upon to form a protection for the steel frame against fire. Four-inch or five-inch slabs, such as are often used, form very little protection, and even where the facing is made of a greater thickness, it should be backed up with brickwork or terra-cotta, so as to protect efficiently the structural steel in case the stone veneer is destroyed. This fireproof backing should be arranged so as to be entirely independent of the facing for support, so that the destruction of the latter would not cause the failure of the protective covering.

If limestone, marble, or granite is used for the exterior, the design should be such that the strength of the structure does not rely upon the stone.

The effects of fire upon marble or limestone exterior walls

were clearly shown in the cases of the Home Life Building and Chicago Athletic Club Building, in both of which extensive rebuilding was necessary. These fires point to the desirability of employing the pure skeleton construction if the use of stone is introduced to secure architectural effects. The stonework would then be free from any load-carrying functions, and the walls would be divided story by story. Damaged material could be replaced without disturbing unaffected areas.

Brickwork.—Brick masonry is usually employed for the body of the exterior walls. These should be of sufficient thickness and rigidity, and built of the best possible materials.

Four inches, or a single thickness of brick, is generally considered an efficient covering for steel members, but 8 ins. is much to be preferred as a minimum, both on account of fire-resistance and protection against corrosion. Cement-mortar should always be used in the best classes of work, especially where it comes in contact with the steel frame.

Terra-cotta.—Ornamental terra-cotta is very extensively employed in exterior wall-construction, either for the ornamental portions only, or for entire façades which are now becoming very common in skeleton-construction buildings. Where a combination of brick and terra-cotta is used, the latter material is generally used for belt-courses, friezes, sills, lintels, and jambs in the main wall surfaces, and very often for cornices, pediments, or balconies which project beyond the building lines.

The admirable weather- and fire-resisting qualities of ornamental terra-cotta were pointed out in Chapter V, and no more satisfactory building material can be employed for exterior-wall construction.

Terra-cotta blocks used in exterior walls should invariably be backed up with brick masonry, and in all possible cases the hollow faces in the rear of the terra-cotta blocks should be well filled with mortar, into which bricks or parts of bricks are to be worked to make the masonry as solid as possible. Cement-mortar only should be used. The brick backing should be anchored to the steel frame, either by hooking-rods or anchors over portions of the frame, or by passing anchors through holes punched in the frame for that purpose. The individual terra-cotta blocks must be anchored to the brick backing or to the steel frame direct. The terra-cotta is usually built up in advance of the backing, one course at a time. Examples of fastening the terra-cotta and backing are given in a following paragraph describing Spandrels.

After setting, the joints should be raked out to a depth of $\frac{3}{4}$ in. and pointed with Portland cement colored to suit the shade of terra-cotta employed.

Party-walls.—Party- or dividing-walls should be absolute barriers against the spread of fire, and weak or thin walls for such locations will be false economy. In many cases the steel frames for large and important structures are placed directly against the walls of the smaller adjacent buildings, or out to the party-lines. In case of conflagration, the smaller building will pull down its own wall, and leave the steelwork of the newer structure exposed. The columns and wall-beams should be efficiently protected by their own wall.

Anchorage.—The proper anchorage of the steel members should be provided for, to tie the beams and girders, and, if necessary, the columns, to the masonry walls. This is often overlooked, and in many specifications nothing whatever is said on this point. All beams and girders should be securely anchored to the brickwork, especially where load-supporting walls are used, as this will largely prevent the collapse of the walls from hot-air explosions, etc.

Thickness of Walls.—For all cities where steel-frame buildings of a fireproof nature are required, the local building ordinances will determine the necessary thickness of the walls, whether of the load- or self-supporting type, or veneer construction.

The New York City building laws require that 12-in. curtain-walls shall be carried at every floor, but 16-in. curtain-walls may be built two stories high without intermediate support. The same law requires the use of 12-in. curtain-walls for the top four stories, or for about 50 ft. down from the roof, and 4-ins. additional thickness for every lower 50 ft. section down to the sidewalk-level. These requirements are so severe as applied to skeleton-construction buildings of great height, that the Board of Examiners, who are empowered to modify the city building laws within certain limits, have often made concessions on this point, but they never permit 12-in. walls for the full height of a building over twelve stories.

Openings in Walls.—All openings in exterior fireproof walls should be provided with efficient fire doors or shutters. This rule applies to *all openings*, and not to doors or windows only. It is the careful attention to minor means of communication that often insures protection in case of emergency. Openings left for the passage of pipes, flues, belts, or shafting, etc., should receive an equal consideration with doors or windows. Fig. 113 shows a fireproof casing for a shaft-opening in a fire wall, as recommended by the Boston Board of Fire Underwriters.

Where practicable, exposed openings in exterior walls should be provided with fire shutters, or with metal or metal-covered frames and wire-glass, as stated in Chapter VII.

Terra-cotta Furring-blocks.—These consist of blocks or slabs placed against the exterior walls to secure insulation against dampness, or to prevent the conduction of heat or cold, or the transmission of sound. They are extensively used in refrigerator or cold-storage buildings and in breweries, to preserve a uniform temperature. In dwellings, stores, and mercantile buildings they are generally used to

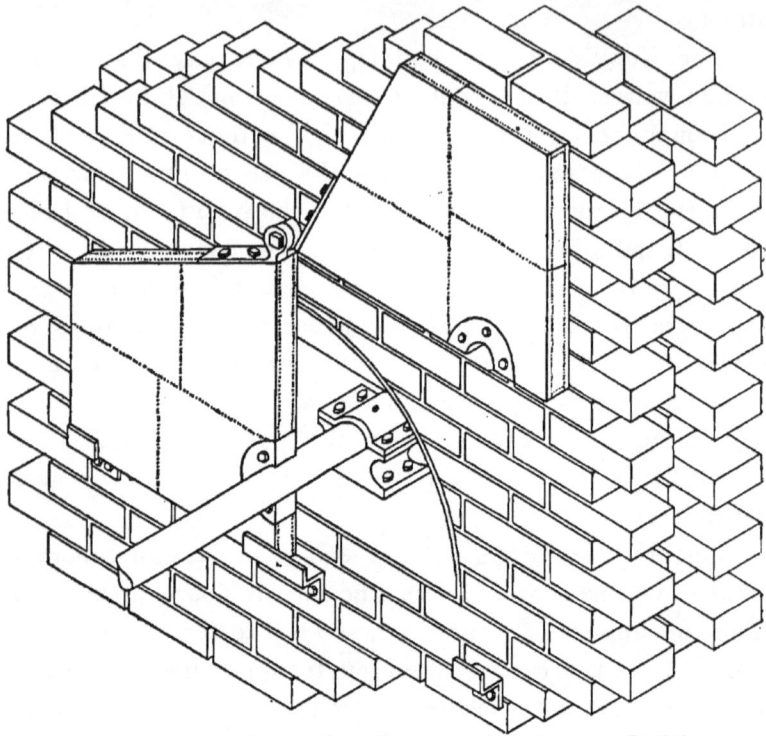

FIG. 113.—Standard Cut-off for Shaft-openings between Buildings.

prevent the penetration of dampness, and in churches, theatres, etc., for acoustic properties.

For stores, offices, and dwellings, " single-thickness " blocks with a 1-in. or 2-in. air-space are usually employed. This form is shown in Fig. 114. Blocks 12 ins. by 16 ins. by 2 ins. weigh 8 lbs. per sq. ft. " Double-thickness " furring-blocks, which are the same as partition blocks, are used from 3 ins. to 6 ins. thick. In breweries or cold-storage houses the walls are sometimes lined with two layers of double-thickness blocks, between which is placed a filling of mineral wool.

FIG. 114. — Single-thickness Terra-cotta Furring-block.

Terra-cotta furring-blocks are not essential for fireproofing, as ordinarily employed, but they are often valuable in preventing the transmission of severe heat, and as their surfaces receive the plaster finish direct, they dispense with the use of lath. They are secured to the walls by setting in gauged mortar, or by using flat-headed nails which are driven into the joints of the brickwork at suitable intervals.

Spandrels.—Spandrels constitute those portions of the exterior walls which lie between the piers and between the window-spaces of the successive stories.

In load-supporting or self-supporting walls, these portions present no especial difficulty, as they are easily cared for by introducing lintel-beams, channels, angles or tees, which rest directly upon the masonry piers, or the support is made through the use of stone lintels, or by the arching of the masonry over the openings.

In veneer or skeleton construction, however, considerable ingenuity is often required to carry the terra-cotta construction properly. The spandrels are often made thinner than the piers surrounding the exterior columns, in order to reduce the loads on the spandrel-beams, as well as to throw the spandrels "in reveal," thus accentuating the piers for architectural effect. In this construction the support of the brickwork and fireproofing, and the proper attachment of the terra-cotta blocks, bring up many problems calling for originality and practical adaptability.

The methods employed can be best described by means of illustrations. Quite a number of examples are given in the author's volume on "Architectural Engineering," with descriptions of the ordinary methods employed in skeleton-construction buildings in Chicago. A few examples only will be here pointed out, referring especially to the fireproofing of the spandrel- or lintel-beams.

Fig. 115 illustrates the spandrel-section and elevation at

the second floor-level of the new " Cable " Building, at the corner of Wabash Avenue and Jackson Boulevard, Chicago.* This section is typical of the detail commonly employed for

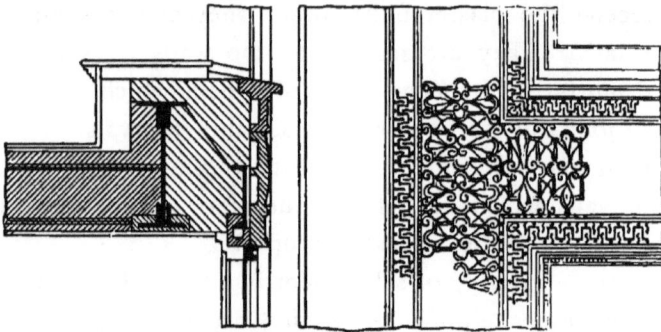

FIG. 115.—Spandrel-section and Elevation at Second Floor—Cable Building.

thin curtain-wall construction. The floor-system and the brick backing are carried by an I-beam and channel, to which latter member is riveted an angle for the support of the terra-cotta lintel and the blocks above.

Fig. 116 shows the spandrel-section at the third floor-level of the same building. In this case the wall is made much

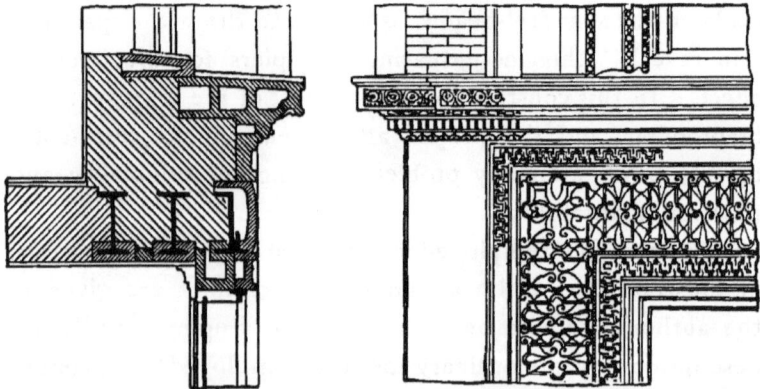

FIG. 116.—Spandrel-section and Elevation at Third Floor—Cable Building.

thicker, to provide a deep reveal for the windows of the stories above.

* Architects, Holabird & Roche, Chicago.

Fig. 117 is a typical section applying to the fourth to tenth floors inclusive. The construction is very similar to

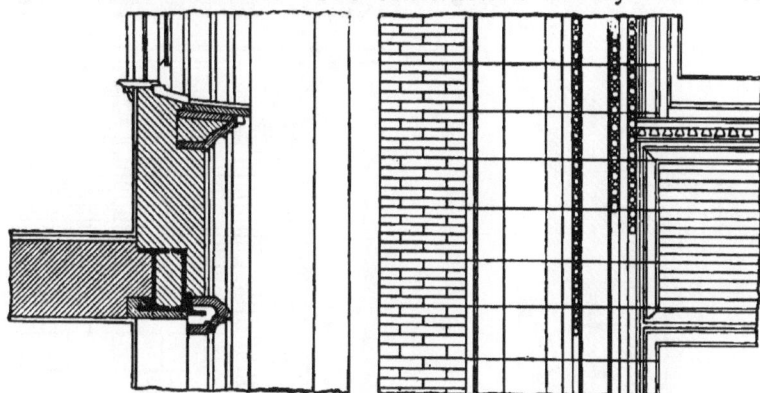

FIG. 117.—Spandrel-section and Elevation at Typical Floor, Cable Building.

that at the second floor-level, except that the spandrel is placed much farther in from the building line, to obtain the deep-set windows.

Fig. 118 is a section and elevation showing the roof and cornice construction.

In designing lintels or spandrels, especial care is necessary to see that thorough protection is secured for the steelwork against possible attack by fire. The lintel protection-pieces should be especially designed to fit their places, and be so arranged as to extend out and join the terra-cotta on the front of the building, thus entirely covering and protecting the lintel- or spandrel-beams.

To accomplish this, hooks and clamps must often be employed in much the same manner as described under the head of Girder Protection in Chapter VIII. To secure the best results, the various sections should be drawn to full size, and the best possible arrangement carefully studied out to secure a stable and efficient covering for the steelwork.

Fig. 119 shows a section through the head of the court-windows in the Cable Building.

Fig. 120 shows the section through an opening in the alley

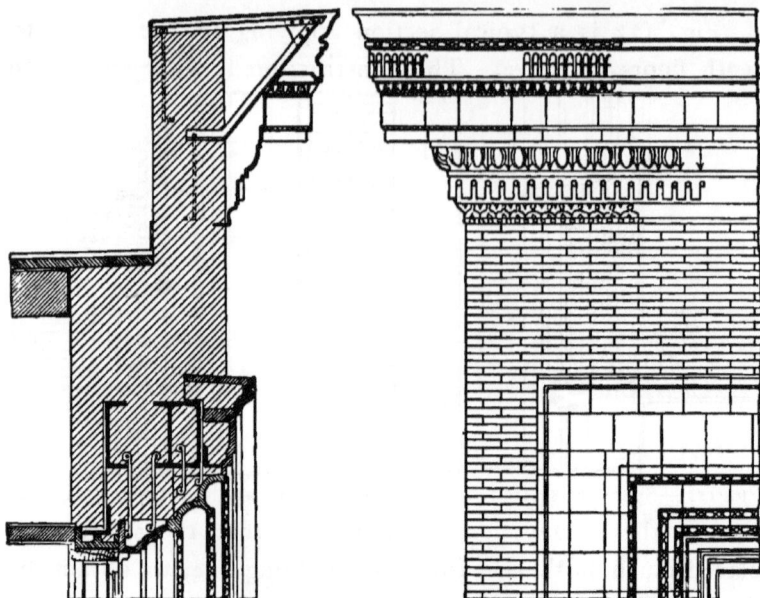

FIG. 118.—Section and Elevation at Roof-cornice—Cable Building.

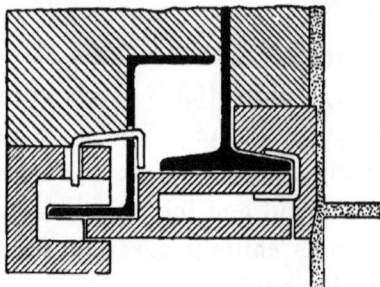

FIG. 119.—Lintel Protection—Court Windows, Cable Building.

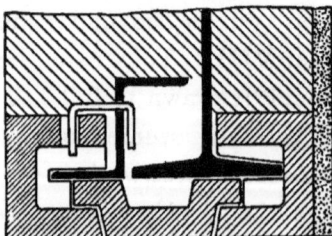

FIG. 120.—Lintel Protection—
Opening at Backing-in Space,
Cable Building.

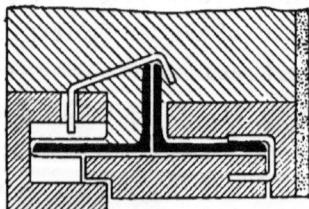

FIG. 121.—Lintel Protection—
Alley Windows, Cable Building.

wall at the same building, and Fig. 121 shows a section through the heads of the windows on the alley elevation.

These sections illustrate most commendable methods of fireproofing the lintel- or spandrel-beams, and the employment of similar details for all spandrel protections will undoubtedly give the best results under severe test conditions.

On the front wall of the Horne Store Building in the Pittsburg fire the spandrels were carried by means of cast-iron lintels which rested on the piers, but which were not fastened to the steel frame. On most floors these lintels came close down to the window-frames, and were thus exposed to the action of the fire. In the fourth story a partial protection was attempted by means of a course of terra-cotta blocks, but these were fastened in place by iron anchors passing through the terra-cotta blocks, so arranged as to be themselves exposed to the fire.

Special constructions such as balconies, bay-windows, etc., all demand individual treatment, depending upon the design. The principles which have already been described must be adapted to the conditions to be met, the one main thought from a fireproofing standpoint being the complete protection of the structural steelwork.

Fig. 122 illustrates the construction employed for the support of the bay-windows in the Great Northern theatre, hotel, and office building in Chicago.

Fig. 123 shows a section through the show-window bulkheads and through the sidewalk of the Cable Building.

Columns in Exterior Walls.—Some architects and engineers have advocated placing the exterior columns entirely within the building, either two or three feet inside of the interior face of the exterior walls, or else so that the fireproof covering of the columns will just join the inside of the walls. The claim made for this construction is that the metal columns will then be under a more even temperature the year round,

that they will be better protected from the moisture or
deteriorating influences of the exterior walls, and that they

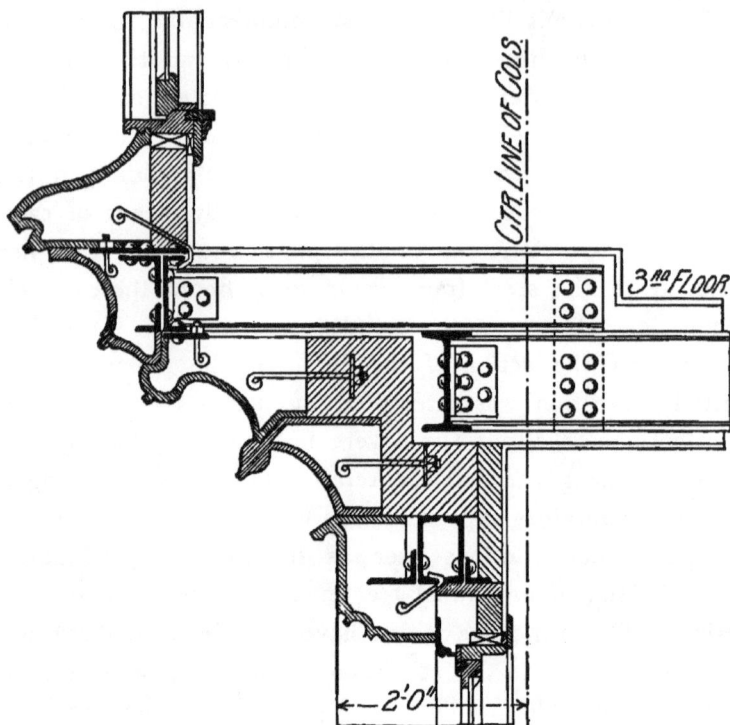

FIG. 122.—Bay-window Construction.   Great Northern Hotel and Theatre
Building.

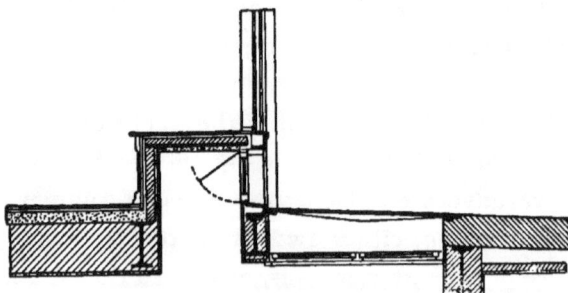

FIG. 123.—Section through Bulkheads and Sidewalk—Cable Building.

will be better protected in case of fire from outside sources.
Greater facility of inspection is also claimed for this method.

In the Havemeyer and St. Paul buildings, New York City, designed by Geo. B. Post, architect, the exterior columns are carried up in this manner clear of the walls, instead of being built into and enclosed by the walls, as is the usual practice in skeleton construction. The principal object sought after in these cases was to avoid embedding the steel-work in the masonry, where inspection would be impossible to determine whether the metal had suffered from the deteriorating influences of moisture or exposure to severe heat. The wall-columns were therefore placed just inside of the inner faces of the piers. Each column was encased with porous terra-cotta tiling 4 ins. thick, entirely surrounding the column, even on the wall side. Between this fireproofing and the outer wall were laid two courses of heavy tarred or asphalted felt, and the space between this felt and the brick-work was filled in solid with cement-mortar.

The author does not believe that this method of construction possesses any advantages from the standpoint of fire-proofing, while the elements of cost, loss of floor space, and eccentricity of loading act as drawbacks to this arrangement. As to inspection, while of undoubted value *if done* at intervals, it is difficult to believe that the average owner would be any more likely to cause the fireproofing of an isolated column to be stripped off for inspection than that of a column embedded in a masonry wall with a terra-cotta casing on the room side.

Columns properly embedded in walls can be made quite as fireproof, if not more so, than when isolated. Columns placed inside the wall-line, as above described, may be better protected against fire from without the building, but from fires within, isolated members must endure severer conditions than those embedded in masonry walls or partitions.

The method generally employed has been to place the column within the wall in such a way as to leave from 4 ins.

to 12 ins. of masonry between it and the outer air, allowing
the remainder of the column to project into the rooms, with

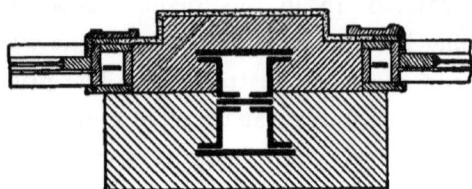

FIG. 124.—Fireproofing of Columns in Exterior Walls—Old Method.

hollow-tile fireproofing on the back and on portions of the
sides. This is shown in Fig. 124.

In later and better examples the fireproofing is carried
entirely around the columns, and the brick or terra-cotta
front is not relied upon as the only external protection, even
though the column be completely surrounded by masonry.

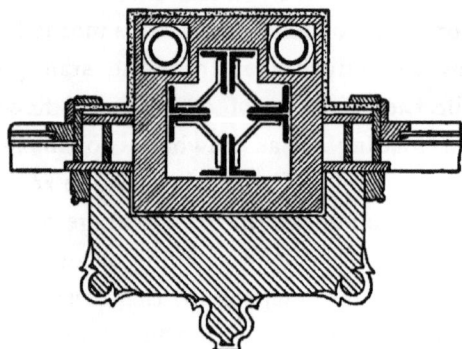

FIG. 125.—Fireproofing of Columns in Exterior Walls—Fisher Building.

Fig. 125 shows the detail employed in the Fisher Building,
Chicago.*

Fig. 126 gives the detail of the corner pier and column
protection at the corner entrance to the first floor of the
Cable Building.

Fig. 127 shows the corner pier and an intermediate pier

* D. H. Burnham & Co., Architects.

at the typical floors of the same building. Compare with
elevation in Fig. 117.

These details constitute a great improvement over the

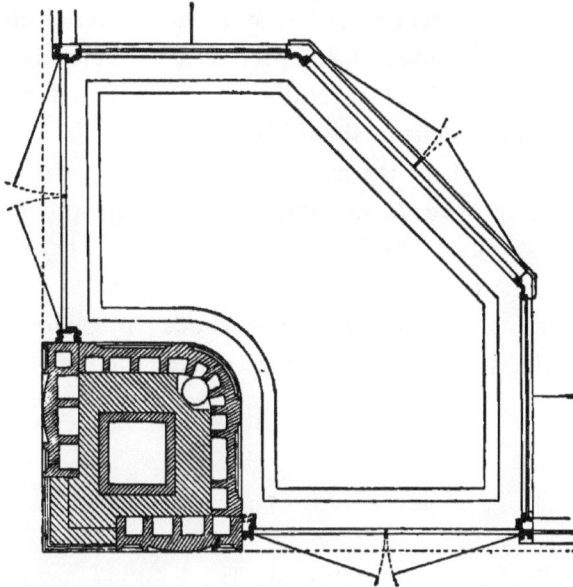

FIG. 126.—Corner Pier and Column Protection—Cable Building.

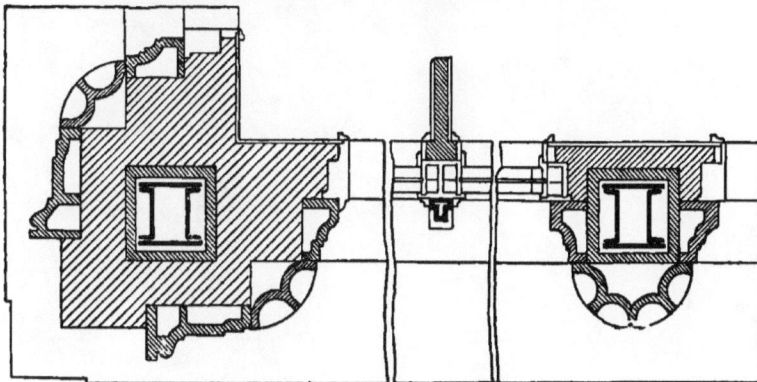

FIG. 127.—Column Protection and Corner Pier—Typical Floor, Cable
Building.

older methods, as the columns in the exterior walls should
be covered with hollow tile just as completely as isolated
columns. In addition, there should be not less than 8 ins.

of brickwork or terra-cotta masonry between the fireproofing and the outer air.   No allowance should be made for stone-work.

In setting the terra-cotta blocks around exterior columns, all of the points noted in Chapter X as to efficient placing should be watched as carefully as for isolated columns. Portland cement-mortar should be used, both for the terra-cotta covering and for the masonry.   If, for any reason, lime-mortar is employed, a coating of some waterproof felt or other impervious material should be placed between the mortar and the metal frame.

Figs. 125, 126, and 127 show methods employed in caring for steam-risers, piping, etc., when carried alongside exterior columns.

CHAPTER XIII.

ROOFS, SUSPENDED CEILINGS, FURRING.

THE principal importance attaching to roof construction in fireproof buildings lies in the duty that the roof has to perform in either excluding external fire or in confining internal fire.

The requirements for excluding external fire are, in most cases, fulfilled by providing a fireproof roof covering, which will act as an efficient shield against sparks or embers from neighboring conflagrations; but where the walls of adjacent structures extend higher than the roof to be constructed, the possibility of falling walls or débris and the attendant destruction are to be considered.

To confine internal fire the roof must act as a perfect barrier to the outburst of flame, for when it is once broken through, the intensity of the fire is rapidly increased by the resulting draught and suction.

In buildings as ordinarily constructed, the under surface of the roof will receive a greater concentration of heat than any other surface in the structure. This is due to the upward rush of flame and heated air by means of vertical courts or light-shafts, or stair- and elevator-wells.

General Methods of Roof Construction.—The roof-framing and the roof-arches are made lighter than the framing and arches for the floors, as most building laws prescribe live-loads to cover snow and wind, etc., which are less than the live-loads required for the floor systems. The dead-load is also less, in that the partition-loads are omitted.

These reductions in loading make it possible to employ terra-cotta in the form of light segmental arches, arches with raised skew-backs, or flat terra-cotta arches of shallower depths than ordinarily employed in floor construction. If a still lighter form is desirable, ceiling tile or book tile are carried on tee-irons, without any arches between the beams or girders. For concrete construction, the same general details are used for roofs as for floors, except that the arch or slab is made shallower and lighter.

For ordinary cases of roof framing the roof-beams are supported by the girders, which run over and are supported by the columns of the top story. The beams and girders must be so àrranged as to give a sufficient pitch to drain the water to the down-spouts, as most conveniently located. Roof construction may be divided into four general forms:

(*A*) The roof and the ceiling of the top story formed by the same construction.

(*B*) A level suspended ceiling hung beneath the roof construction.

(*C*) Pitched roofs.

(*D*) Mansard roofs.

Roof and Ceiling Combined.—This type of roof is generally limited to warehouses or manufacturing buildings, where a flat suspended ceiling is not necessary for appearance. Unless the roof-beams are made perfectly level, as for a floor, with the roof pitch made up in concrete or other filling over the arches, the result is an irregular pitched ceiling, due to the slope of the roof-beams for draining the roof-surface. This is not objectionable in storage or manufacturing buildings, and these forms are generally employed in such structures.

Great care is necessary to secure the thorough fireproofing

of all beams and girders, and the recommendations made in Chapter VIII as to girder protection should be carefully followed.

The roof-arches may be made either of terra-cotta, flat or segmental, or of concrete arches or slabs. Fig. 128 shows the roof construction employed in the U. S. Public Building at

FIG. 128.—Terra-cotta Roof Construction—U. S. Public Building, Portland, Oregon.

Portland, Oregon, consisting of flat terra-cotta arches with concrete filling, over which are placed five layers of best asphalt roofing-felt, to serve as a waterproof covering. The finished roof-surface is made of 1¼ in. solid flat tile, embedded in cement.

Fig. 129 illustrates the concrete and expanded metal roof used in the U. S. Public Building at San Francisco, Cal.

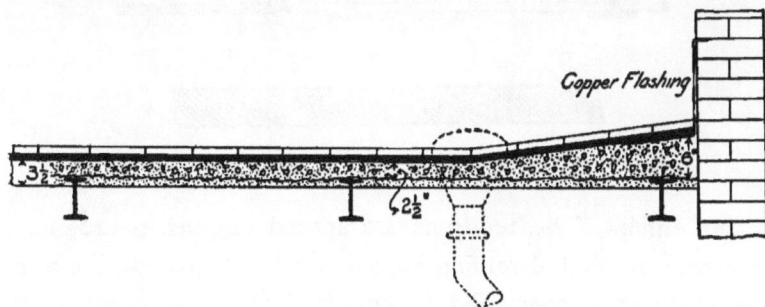

Copper Flashing

FIG. 129.—Concrete Roof Construction—U. S. Public Building, San Francisco.

No. 18 B, W. G. expanded metal is laid over the roof-beams, upon which is constructed a concrete plate 3½ ins. thick—2¼ ins. above the expanded metal, and 1 in. below it. The beams are protected by terra-cotta shoe-blocks. The roof surface is made of five layers of asphalt roofing felt and $1\frac{1}{4}$ in. solid flat tile, embedded in cement.

Form (*A*) is sometimes used by making the roof framing and arches level, as in regular floor construction, and then building up with terra-cotta to give the required pitch above.

This is often done where additional stories are to be erected at some later date, the top floor meanwhile serving as a temporary roof. There is no particular objection to building up in this manner, from either a structural or a fireproofing standpoint, but the condensation on the ceiling-surface, caused by ordinary differences in external and internal temperatures, will, even with the hollow tile, constitute a serious difficulty.

A lighter but less efficient roof and ceiling construction is made by placing the beams sufficiently close to carry 3-in. or 4-in. tee-irons, upon which are placed roofing tile, sometimes called book tile. See Fig. 130, (*A*). These blocks are commonly 12 ins. wide, 14 ins. to 20 ins. long, and 3 ins. to

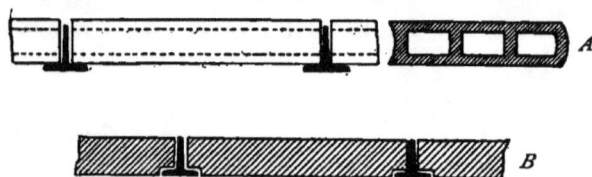

FIG. 130.—Terra-cotta Roofing Tile.

4 ins. thick. The tee-irons are spaced $14\frac{1}{2}$ ins. to $20\frac{1}{2}$ ins. centres. Rabbeted roofing-blocks, see Fig. 130, (*B*), are also largely used, either solid or hollow, but for approximately horizontal load-bearing surfaces the rabbeted blocks are considerably weaker than the book tile. Either style is stronger when made hollow than when made solid, and hollow blocks should be used in all places except where the attachments of flashings require solid blocks. This detail of roof is used for lighter systems of roof construction, but it is

generally considered more expensive than the ordinary roof-arches. The weight of the tee-irons forms a large item of expense.

As against external fire, a tee-iron and book-tile roof will give sufficient protection, but it will not offer the resistance to shock or load possessed by arch construction. This is important where walls or portions of walls of adjacent buildings are liable to fall upon the roof-surface during a conflagration.

Considering internal fire, this construction provides no adequate protection for the under sides of the tees. Where book tiles are used the tees project below the bottom surfaces of the terra-cotta blocks, and a very thick coat of plaster on the terra-cotta is necessary to cover the tees even with a very thin coating. Where rabbeted tiles are employed, the plaster coating over the tee is thicker, but as it has little or no bond to the iron surface, the protection is only nominal. This difficulty of satisfactorily fireproofing the tee-irons renders this type of roof unsuitable for structures intended to be of the best fire-resisting construction.

Another objection lies in its comparative thinness, as this method does not provide a good insulation against changes in temperature. The temperature of the spaces under the roof will be easily affected by outside changes, and condensation will occur under even slight differences in temperature.

Roofs with Suspended Ceilings.—Where appearances must be considered, as in mercantile or office buildings, a level ceiling must be had under the sloping roof-surface. This is generally accomplished by suspending a light ceiling construction from the roof-beams.

This does not change the roof construction itself in any way from the preceding form. The same general details are used, with the addition of the suspended ceiling for the sake of appearances. Fig. 131 shows the roof and ceiling construction employed in the new U. S. Appraisers' warehouse, New York City, in which the roof consists of a 3-in. plate of concrete, with expanded metal embedded therein, over which are laid two layers of asphalt, $\frac{1}{2}$ in. thick each, for the finished roof-surface. The suspended ceiling is made of 2-in. by 2-in. tees, suspended under each beam, upon which rest $1\frac{1}{2}$-in. by $1\frac{1}{2}$-in. tees, spaced 16 ins. centres, to receive the expanded

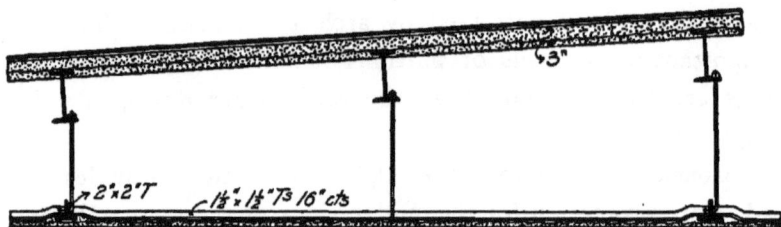

FIG. 131.—Roof and Ceiling Construction—U. S. Appraisers' Warehouse, New York City.

metal lath and plaster. In this particular building asbestic plaster was used throughout.

Other forms of suspended ceilings will be described in a later paragraph.

Where a suspended ceiling is used, all spaces between the ceiling and the roof should be made inaccessible, and no pipes or other mechanical features should be located in such voids. This is to make sure that no one may have cause to visit these places and leave the means of communication open. All stair-wells, skylights, or light-courts which may extend up through the ceiling to the roof-level should be thoroughly ceiled up between the ceiling and the roof by means of vertical terra-cotta or other fireproof partitions. These should extend entirely around the openings.

In using suspended ceilings under roofs or in other locations in buildings, the thorough fireproofing of the steel members over the ceiling should never be omitted, however thorough the ceiling construction may be made. In this respect the roof shown in Fig. 131 is open to criticism. It is not safe to assume that any ordinary form of suspended ceiling will successfully resist a long-continued or very severe fire. Metal lath and plaster or thin terra-cotta ceiling-blocks supported on light tee-irons have been repeatedly demonstrated insufficient for resisting severe conditions. It is also to be remembered that the roof, or the ceiling of the top story, will receive the severest test by heat of any ceiling or floor in the building, provided any stair-, elevator-, or light-shafts exist by which the fire may travel upward. Under these conditions, the ceiling must be considered as just so much added protection for the roof, and not as a substitute for adequate fireproofing of the roof itself.

It was pointed out in Chapter III that the roof-beams of the Horne Store Building in Pittsburg, and the columns between the suspended ceiling and the roof were entirely unprotected by fireproofing. On the failure of the light ceiling under the severe heat, the roof-beams and the beams under the water-tank soon gave way. On this point the appraisers' report recommended as follows:

"That all portions of the structure between the ceiling and the roof should be well covered, as the heat always goes upward. In this case the intense heat of all the floors in the Store Building rushed up the light-shaft, elevator and staircase openings, and the top portion, having no fireproofing, simply melted and let this immense tank, that was placed on the top of the building, plow its way through the several stories, each adding, as they went, their own weight and contents, and all went to the basement."

Pitched Roofs.—Pitched roofs are employed in a great variety of forms, and under widely varying conditions, but the most common form is that made necessary by the use of roof-trusses. This method commonly applies to theatres, halls, armories, churches, and other buildings of a public nature, in which large areas are required to be covered by roof-trusses without the aid of interior columns. The trusses usually support the rafters, which, in turn, carry the purlins placed close enough together to receive the roof-covering.

The attention bestowed upon the fireproofing of these various members should be in direct proportion to the importance of the service rendered. This means that the roof-trusses themselves should receive the greatest consideration, next to which come the rafters and then the purlins; for if the trusses fail, the entire structure, or roof portion at least, will suffer almost complete destruction.

Few definite rules can be laid down for the successful fireproofing of truss members. The details finally adopted will depend largely upon the specific considerations to be met and the ingenuity with which the difficulties may be overcome. In general, it may be said that all truss members and rafters should be surrounded by complete envelopes of terra-cotta, which must be securely held in position by wires or other metal clamps or fastenings. The most feasible way is to apply terra-cotta coverings in the same manner as described for girder protection in Chapter VIII; that is, the top and bottom protection members are held in place by concealed and protected metal straps, while the side pieces may be secured by wedging, interlocking, or tying in position with wires. When the best possible terra-cotta envelope has been secured, the whole member should then be wrapped and wired with some wire or metal lath, over which is to be placed a thick coating of mortar. For this purpose, no better mortar or plaster could be applied than " fire-mortar," which consists of a fire-clay, without lime or cement. This could be spread on to a thickness of $\frac{3}{4}$ in. and allowed to dry in place. Fig. 132 indicates the methods which may be employed.

For the roof-surface, details similar to those described for flat roofs may be employed, with either inclined arches sprung from rafter to rafter, or book or roofing tiles, laid between the purlins. But as these inclined surfaces are usually slated, requiring suitable nailing-surfaces, the book or roofing tiles are usually made of solid porous terra-cotta. These are made

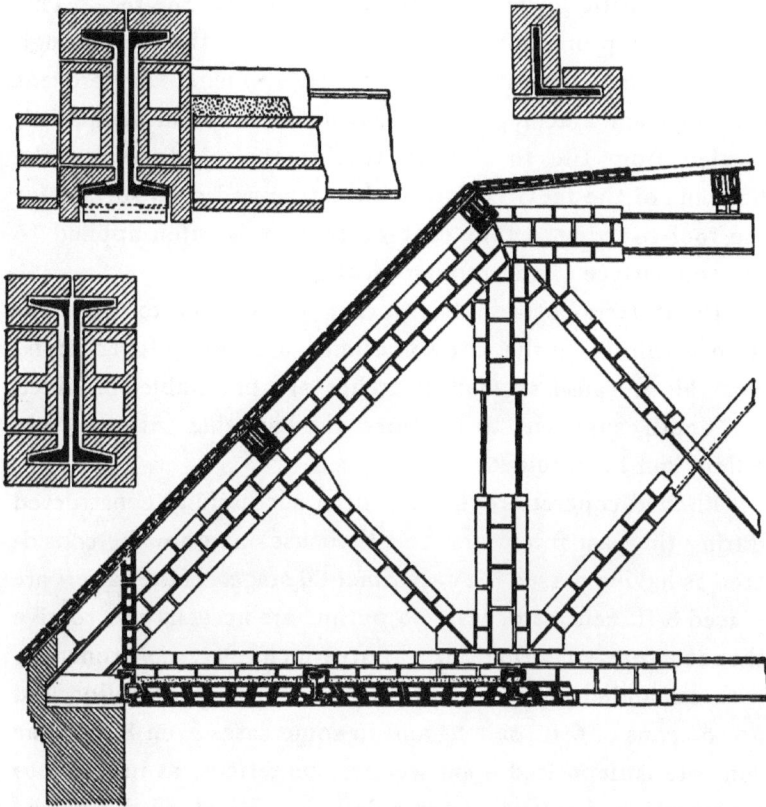

FIG. 132.—Fireproofing of Roof-trusses.

in lengths up to 24 ins., but too much confidence should not be placed in the strength when made in such long lengths. To guard against the possibility of failure by the breaking of such long blocks, especially in armory roofs, etc., where heavy vibration is liable to occur, the best work requires the whole under surface of the roof to be covered with a coarse wire netting of 2-in. mesh. This, when plastered, gives a finished effect to the ceiling.

For use between purlins, rabbeted tiles are much to be preferred to book tiles, for where the under side of the roofing-block is rabbeted, it allows the block to drop as low as, or possibly a little lower than, the under side of the tee. This gives more protection to the under side of the tee through the additional mortar which may be applied. To prevent shrinkage and cracking, strips of metal lath should be applied, stapled from tile to tile, to receive the plaster over the bottoms of the tees. To provide a true and even surface for the roof-covering, a thin coat of cement is often applied to the top surface of the roofing-blocks.

Light terra-cotta arches sprung from rafter to rafter, or from purlin to purlin, are sometimes used for pitched roofs. For this purpose segmental arches are preferable, provided the depth, sufficient to reinforce the haunches with concrete filling, can be obtained.

Pitched concrete roofs have been successfully constructed during the past few years, and their use may now be considered as having passed the experimental stage. If the rafters are placed 8 ft. centres or less, no purlins are necessary to receive the concrete construction. Three-inch plates of concrete, with expanded metal embedded, have been successfully used up to spans of 6 ft. or 7 ft. and in some cases even 8 ft. The concrete is deposited upon wooden centerings, as in the floor construction, and the upper side is smoothed off during the setting, and then is floated smooth and straight to receive the roof-covering. Slates or tiles may be nailed directly to the concrete, without the use of wooden nailing-strips. The concrete holds the nails nearly as well as does wood. For roofs where a nailing-surface is required, the cinders used are screened through a 1-in. mesh.

For concrete roofs, the rafters and purlins may be fireproofed by surrounding them with the regular concrete composition at the time the roof-slabs are formed, and truss members may be wrapped with expanded metal, to receive a plaster coating. But terra-cotta protection is much more satisfactory.

Mansard Roofs.—The inclined members for mansard roofs may either be the only supports for the roof-surface, in which case roofing tile, partition-blocks, or even terra-cotta arches may be laid between them, or the inclined members may support purlins, as in pitched roofs. The same general details of construction are employed as for pitched roofs, but as the principal members of the mansard usually form the supports for the horizontal roof above, the covering or fireproofing of the supporting rafters should be considered as important as that of the trusses or columns.

Attic Spaces.—An attic or storage place is frequently provided between the ceiling of the top story and the roof. In this case the attic floor acts as a ceiling and as a floor at the same time, but the construction is generally made lighter than for the regular floors, due to the reduced loads which it is intended to carry. The clear attic space varies according to circumstances, but it is frequently made 3 ft. to 4 ft. high at the lowest portions, running up to 6 ft. or 7 ft. at the highest points, due to the roof pitch.

In some instances the roof has been supported from the attic floors by means of struts, in which case the attic floor receives the roof-loads, as well as its own loads.

Fig. 133 illustrates the roof construction in the Fisher Building, Chicago. The attic floor is made of flat terra-cotta arches, while the roof is constructed of tee-bars and book tiles, with the supporting beams and girders thoroughly fireproofed. Over the book tiles is spread a layer of cement, on which is placed a six-ply tar and gravel roof.

For a fireproof building, or indeed for any class of structure, attic spaces should be studiously avoided where possible. Such unfrequented spaces are very liable to be stored with rubbish or light materials, which are extremely combustible; and if fires are once started in these areas, they are most

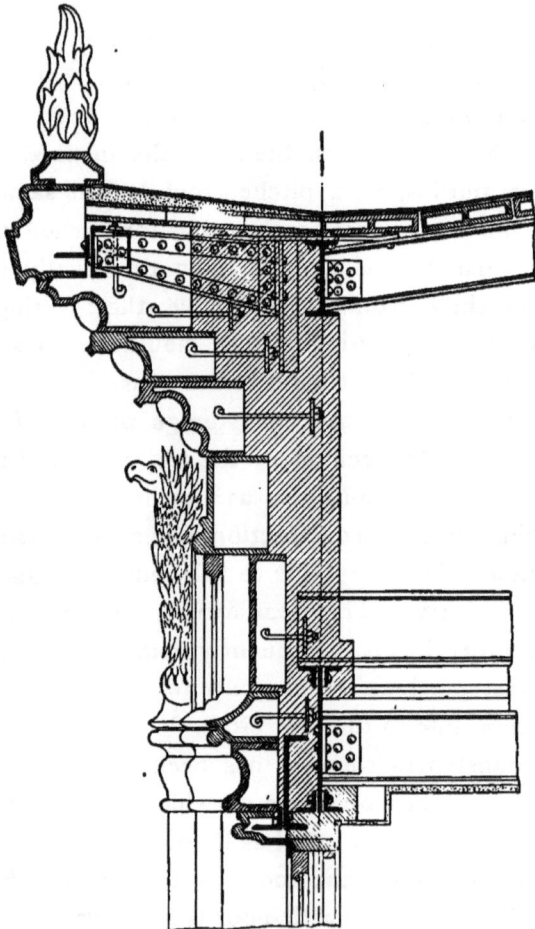

FIG. 133.—Roof and Cornice Construction—Fisher Building.

inaccessible, and likely to be little noticed until of considerable magnitude.

Suspended Ceilings.—Various forms of light suspended ceilings have already been described in connection with concrete and composition floor systems, and ceilings of large areas only will here be considered. These may be arranged to support 2-in., 3-in., or 4-in. ceiling tile of terra-cotta, or simply metal lath and plaster.

Fig. 134 illustrates a suspended ceiling made of $1\frac{3}{4}$ in. by $\frac{3}{16}$ in. flats, spaced 16 ins. centres. These are hung from the roof-beams by means of 1-in. by $\frac{3}{16}$-in. clamps or hangers,

FIG. 134.—Suspended Ceiling Construction.

which are split to clamp the lower flanges of the beams, and bent at the lower ends to provide seats for the ceiling-bars. These bars are laced rigidly in a vertical position, after which stiffened wire lath is wired to the under side of the bars, the stiffening rods in the lath crossing the bars at right angles. This construction weighs from 10 to 12 lbs. per sq. ft., plastered.

Another similar method is shown in Fig. 135. The hangers are spaced 4 ft. centres and are made of 1-in. by $\frac{1}{4}$-in. bars, bent over the lower flanges of the beams, and held in place by $\frac{3}{16}$-in. diam. hooks. If the hangers are quite long, extension-pieces are used, bolted to the clamp portion of the hanger. Flat bars $1\frac{1}{2}$ in. by $\frac{3}{16}$ in., spaced 4 ft. centres, are then bolted to the hangers. These bars are perforated at intervals of 12 ins. centres by rectangular holes $\frac{3}{4}$ in. by $\frac{3}{8}$ in., through which pass small $\frac{3}{4}$-in. channels for the support of the metal lath and plaster. Where these ceilings

are suspended below terra-cotta arches, toggle-bolts are used
for the support of the hangers.

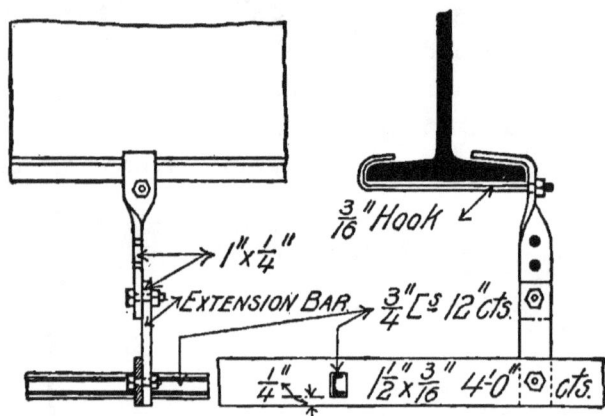

FIG. 135.—Suspended Ceiling Construction.

Fig. 136 shows a very satisfactory detail for a suspended
ceiling made of light tee-irons.   The hangers, spaced 5 ft. to
6 ft. centres, are made of $2\frac{1}{2}$-in. by $\frac{1}{4}$-in. bars, bolted between

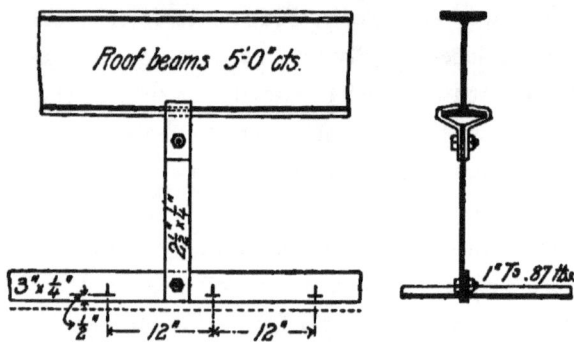

FIG. 136.—Suspended Ceiling Construction.

two clamps, which clasp the lower beam-flanges.   To the
hangers are bolted 3-in. by $\frac{1}{4}$-in. bars, which are punched
every 12 ins. to receive 1-in by 1-in. tees, weighing 0.87 lb.
per ft.   These receive the metal lath and plaster, or if ceiling
tiles are desired, a larger size of tee may be used for the
support of these blocks.   For the splicing of the 1-in. tees,

a sheet-iron clamp, 6 ins. long, is wrapped closely about the tee-flanges. When hammered tight, this makes a sufficiently rigid splice.

In some fireproof buildings the roof and ceiling have both been made with hollow-tile arches, each of independent construction, but lighter than the floors. This is not necessary, either for strength, appearance, or effectiveness. If the roof is made of fireproof arches or plates, with all supporting members carefully protected, a good suspended ceiling is all that is necessary to fill the requirements.

Roofing and Ceiling Tile.—Terra-cotta blocks used for roofing and ceiling purposes are made in three distinct patterns—book tiles, which have two segmental edges, thus making them of the form of a book; roofing tiles, which are plain rectangular blocks, and rabbeted blocks, with rabbeted edge on two sides, to allow the blocks to set down between the tee-supports, as shown in previous Fig. 130, (*B*).

Book tiles are usually made of porous material, 10 ins. or 12 ins. wide, and from 18 ins. to 24 ins. long. In thickness they vary from 2½ ins. to 4 ins. The weight of book tiles 3 ins. thick will average about 12 lbs. per sq. ft.

Roofing tiles, or ceiling tiles without rabbeted edges, are made from 8 ins. to 12 ins. in width, 1 in. to 4 ins. in thickness, and in lengths varying from 10 ins. to 24 ins. They are made of both porous and hard-burned material, but hard tiles should always be scored to receive the plaster finish properly. The weights per square foot will average about 12 lbs. per sq. ft. for 2-in. blocks, and 14 lbs. per sq. ft. for 3-in. blocks.

Rabbeted roofing or ceiling tiles are made both solid and hollow, and of either porous or hard-burned material. The width is usually 12 ins., with the length varying from 10 ins. to 24 ins. The common thicknesses are 2, 2½, 3, and 4 ins.

Two-inch blocks will weigh about 12 lbs. per sq. ft., 3-in. 14 lbs., and 4-in. 18 lbs.

Roof-coverings.—Ordinary roof-coverings for fireproof buildings are made of:

1. Tar and gravel.
2. Asphalt and gravel.
3. Vitrified (glazed) tiles.
4. Brick.
5. Slate tiles.

In price and in efficiency under long-continued wear and severe conditions these will rank in about the order named.

The cheapest roof-covering in common service is that of tar and gravel. This is made by laying tar-paper, 5- or 6-ply, on the concrete filling which is placed over the arches. Each sheet is thoroughly mopped as laid, after which roofing-gravel is spread over the entire surface. Asphalt roofs are constructed in the same manner, except that asphalt-felting, 5-ply, is used, mopped with hot asphalt during laying.

Vitrified tiles, about 8 ins. by 8 ins. by $1\frac{1}{2}$ ins., are laid on the concrete filling in elastic cement. Slate tiles, usually 12 ins. square by about 1 in. thick, are laid in the same manner. Both kinds make an excellent fireproof roof, the slate tiles being somewhat more expensive.

Brick roofs are very serviceable and thoroughly fireproof but the question of weight has prevented a more general use of this material. If economy need not be especially considered, their use is to be highly recommended. Five-ply tar-paper is first laid over the concrete leveling, and the bricks are then placed on edge in Portland cement. The cost is about the same as for vitrified tile.

Tin or copper roofs should only be used over wood, as over terra-cotta or concrete the sheets of tin or copper will rust out in a few years.

Pent-houses.—In cases where stair-wells or elevator-shafts are not surrounded by brick walls, or where masonry walls are used but are not carried up through the roof, pent- or roof-houses are constructed of steel framework and terra-cotta blocks or concrete.   Either flat or double-pitched roofs are used, with or without skylights, as may be required.

The steel framework is usually made of vertical 3-in. tee-irons, spaced about 18 ins. centres, with angles at the corners and at door- or window-openings.   A horizontal frame or plate made of angles or channels surrounds the top, and pro- vides a seat for the skylights or roof tees or beams.

In pent-houses surrounding elevator-shafts, where the sheave-beams are connected to the pent-house frame, thus causing constant vibration, the flanges of the tee-uprights should be placed on the inside.   The blocks are then placed from the outside, and if they become loose from vibration, the tees will prevent their falling inward, while the outside covering of tin or copper will keep the tiles hugged in place.

The same construction is often employed for "skylight curbs," where skylights are placed several feet above the roof-level.

Wire glass should always be used for all skylights, or for window-openings to pent-houses.

Metal Furring.—The introduction of the various forms of metal lathing has greatly developed the use of furring to produce architectural forms in interior decoration.   Very elaborate effects are now easily produced by the aid of metal furring, lathing, and plaster, where formerly such effects were only possible in very heavy and very expensive construction. Cornices, coves, false-beams, arches, and domes, can now be readily and economically constructed with a false-work of furring, where previously the same effects in massive con- struction would have been difficult, heavy, and expensive. A great quantity of this false-work now enters into nearly all large buildings, and the uses to which metal furring are applied are as numerous as the conditions which call forth its use. The furring is always of a sham nature, representing forms of construction which do not exist. It is never employed to carry loads of any magnitude, and generally nothing but the weight of the plaster or mosaic finish.

For cornices, false-beams, etc., furring members of light bars, angles, or channels, are attached to the walls by means of nails, staples, toggle-bolts, etc., or to the steel beams by

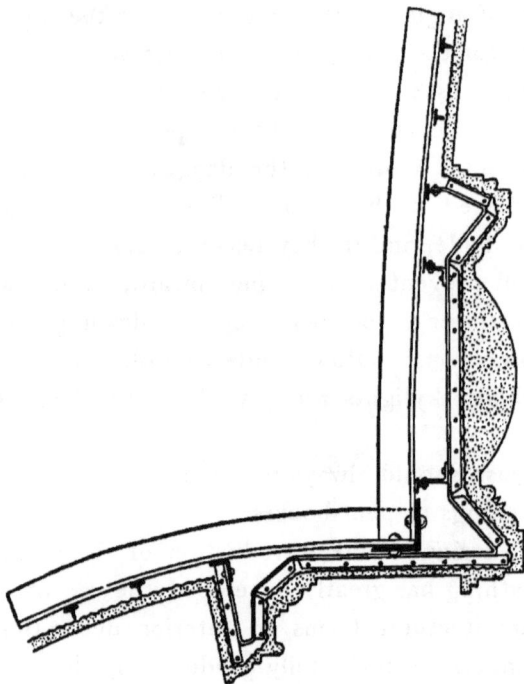

Fig. 137.—Cornice Furring—House of Representatives, Mass. State House.

means of hangers, stiffening rods, etc. These furring pieces are bent or shaped to the required outline, so that for plaster work the approximate outline of the metal lath which is attached to the furring members shall not be more than $1\frac{1}{2}$ ins. distant from the finished plaster-line. Fig. 137 shows an elaborate cornice employed in the Massachusetts State House extension.

Flat ceilings are often employed for architectural effect where the ceilings in rooms, hallways, or vestibules are required to be at a lower level than the actual floor construction. Flat ceilings are also employed to conceal pipes or vents which may tend to disfigure the apartment. Light tees or angles are generally used for this purpose, spaced about 12 ins. centres. For spans not exceeding 5 ft., these may be run into the walls for support, but in longer spaces, intermediate hangers are necessary.

For larger arches or domed ceilings, structural members must be provided at intervals to carry the lighter iron falsework, metal lathing and plaster. These usually consist of tee- or angle-irons bent or shaped to the proper outlines, and spaced at intervals of from 4 ft. to 6 ft. Such members are placed about 2 ins. back from the finished plaster-line. Light tee-irons, 1 in. or $1\frac{1}{4}$ in., are then clipped to the supporting framework at intervals of 12 ins. These receive the metal lathing, which in turn takes the plaster finish.

CHAPTER XIV.

EQUIPMENT FOR FIRE-RESISTANCE.

IN Chapter III it was pointed out that partly through misrepresentation and partly through ignorance in regard to the limitations of perfect construction, the general public has been led to expect little short of infallibility in fireproof design. With the erection of a steel framework, and its covering of terra-cotta, concrete, or other incombustible material, non-professional judgment has been too prone to assume the complete protection from damage by fire of the building and its contents. That these expectations have not been realized, even in the very best class of structures, has been sufficiently emphasized in previous chapters.

Fire protection, viewed in its broadest light, should not only mean that the walls and floors will retain their stability under fire-test, but the interests of the owner and the insurance companies also demand that the integrity of the structure as a whole should be well preserved after a conflagration, and the repair needed reduced to a minimum.

Again, adequate fire protection does not concern the building alone. Damage to the contents is not eliminated by simply providing an incombustible structure for their reception. Fire may spread quickly and sweep through a building, with an ultimate heavy loss on stock or contents, and comparatively little damage to the structure itself.

Another difficulty in fire protection is found in the inability of the city fire departments, through circumstances largely beyond their control, to handle fires once started. This particularly applies to buildings of great height, where even the most modern apparatus of the fire departments is wholly ineffective, and where, unless auxiliary fire protection is provided within the structure itself, the fire when once established must be left to burn itself out above the level of the range of the fire departments' apparatus.

These considerations show the necessity of providing efficient equipment to protect and strengthen even the best fireproof construction, which, if left unguarded by such additional devices, may insure but a partial salvage and utterly fail to secure the contents against almost total loss in case of severe conflagration.

In contending with fire, an ounce of prevention may save pounds of loss. Except in the cases of severe exposure fires, conflagrations usually result from very small causes. If discovered in the incipient stage, either by the use of automatic alarms or by other mechanical devices, such as automatic sprinklers, etc., the danger of the spread of the flames and the incumbent destruction may be entirely obviated.

The principal auxiliary aids in caring for fire through equipment consist of:

Automatic alarms;

Automatic sprinklers;

Open sprinklers;

Standpipes and hose-reels, etc;

Accessibility and elevator service;

Inspection of apparatus provided;

Discipline of employees.

Automatic Fire-alarm Systems.—Automatic fire-alarm systems depend upon thermostats, or heat-detectors, for the indication of an abnormal or dangerous rise in temperature. The thermostat causes an electric circuit to be completed or broken, which in turn causes an alarm to be sounded, indicating to the proper parties that either a dangerous temperature exists at some particular location or that a fire has started.

An improved form of thermostat consists of a compound metallic spring. This is contained within a small metal case, about 2½ ins. in diameter. The bottom or exposed surface of the case is perforated, to admit the heated air, and an opening is also provided through the top, to permit a free circulation of air over the spring, and to avoid the possibility of having a cushion of cold air within the thermostat. This is to render the device as sensitive as possible.

One end of the metallic spring is connected to one wire of the circuit, while the other end is free but almost touching a set screw, which is connected to the other wire of the circuit. The thermostat can be set to any degree of sensitiveness by making the spring so that it will expand a greater or less distance, thus bringing it into contact with the screw. It may be made so sensitive that an increase of a few degrees of temperature is sufficient to make the contact, while the heat of the breath will readily operate one if set within a delicate range. The thermostats are usually set to operate under temperatures from 35 degrees to 40 degrees above the normal maximum temperature to be expected in the premises. The operating temperatures are determined by the automatic fire-alarm company in conjunction with the Board of Fire Underwriters.

The increase in temperature to the predetermined degree causes the spring to expand and make an electrical contact, thereby operating an electro-magnet within a signal-box located within the premises. This transmits an alarm, indicating the particular box number, and hence the building and floor in the building, to a central station, where the alarm is then given to the city fire department. This method of registering automatically the particular building and floor in the building, forms a valuable feature in locating the fire or unusual temperature.

A system has been devised, and is now in use, whereby broken wires do not disable the protective system. This is accomplished by using the duplex system of wiring, combining a metallic with an earth circuit.

The thermostats are always installed upon the ceilings, or at the highest points in open spaces such as well-holes, etc. They are also placed in all closets or spaces where fire could possibly originate. They are usually placed every 15 ft. on open ceilings, or one for every 225 sq. ft. In installing, all of the wires may be concealed within insulated piping, which is covered and hidden by the plastering. Nothing need be visible save the thermostats themselves, and when these are painted the same color as the ceiling they are hardly noticeable.

All circuits and apparatus are tested once a month by the fire-alarm company, and once in every six months in conjunction with the Board of Fire Underwriters.

With the thermostats, a system of " Manuals " or pull-boxes is now generally installed, so that in case of fire the occupants of the building need not wait for the fire to reach a temperature sufficient to cause the action of the thermostat, but upon discovery an alarm may be sounded at once to the central station. These manuals are also valuable where fire is seen in nearby property.

The great value of an automatic fire-alarm system must be evident from the foregoing description. In case of fire, or undesirably high temperatures, these " watchmen who never sleep " serve to notify at once the proper parties, and by these means they constitute the greatest possible protection to all classes of structures. They are now extensively used in mercantile and office buildings, hotels, schools, and dwellings, even though these structures be of the most approved fire-resisting design and construction.

The value of thermostats has been amply demonstrated in many cases. But what they actually seem to accomplish must not alone be considered. The possibility of what might take place if they were not employed must also be taken into account as affecting both the structure itself and the contents.

Automatic Sprinklers.—The practical value of automatic sprinklers is most strikingly illustrated by the record of the " Boston Manufacturers' Mutual Company " for the year 1897. Fire-insurance risks were carried in that year to an amount exceeding $100,000,000.00. These policies were carried on factories and retail and wholesale store buildings, all of which were equipped with automatic sprinkler systems. The losses for the entire year were less than one cent on each one hundred dollars of insurance carried.

Wherever combustible merchandise constitutes the contents of a building, automatic sprinklers are of great value, and in buildings of a height so great as to make the upper floors difficult of access, especially if containing large areas and very combustible contents, sprinklers constitute the best protection obtainable.   •

From the underwriters' standpoint, the insurance companies feel that they can make large allowances on premiums where sprinkler systems are installed, because they know that this equipment will prevent serious damage either to the stock or to the building, from either internal or external attack by fire. The Boston Board of Fire Underwriters allows a rebate of 40 per cent. on the total amount of the premium, if automatic sprinkler systems are installed with watchman service, or with automatic alarms. This allowance applies to warehouses and store and office buildings alike— no greater inducement being offered by the insurance companies on office buildings than on manufacturing or warehouse buildings,

The use of sprinkler systems is hardly to be expected in office buildings, as there is usually not enough combustible material in either the building itself, or in the contents, to warrant the expenditure by the owner. Their installation, however, in the upper floors, at least, of very tall structures, would undoubtedly prevent a repetition of the damage which occurred in the case of the Home Life Insurance Company's Building. For warehouses, factories, and wholesale and retail store buildings, automatic sprinklers should invariably be installed, even in buildings most perfectly designed to resist fire.

An automatic sprinkler is a device for distributing water by means of a valve which is arranged to open under the action of heat, as from a fire which it is intended to extinguish. The distribution of water which results from properly located sprinklers, occurs in the form of a rain of jets or drops, and is sufficient to drench almost any inflammable stock beyond the point of ignition. The distribution is also economical, as the water is more evenly applied than from a nozzle attached to a fire-hose, and the source is directly above the fire.

The requisites for efficient automatic sprinkler protection are:

1. The building must be open in construction, and free from concealed spaces where water thrown from sprinklers cannot penetrate.

2. The sprinklers must be located so that their distribution will cover all parts of the premises.

3. The sprinkler-piping must be of sufficient capacity and must be under water-pressure at all times.

4. The available water-supply must, at all times, be of sufficient quantity and pressure.

All of these conditions are essential to obtain proper automatic sprinkler protection.

Sprinkler-heads.—These consist of sealed orifices which are arranged to open under a predetermined temperature. The opening, for which the standard size is $\frac{3}{8}$ in. diameter, allows the water to fall upon a deflector plate which causes the spray or distribution of the water. A great number of patent heads are on the market.

Each automatic sprinkler should have an unobstructed outlet of such size and form that with 5 lbs. pressure, maintained at the sprinkler, it will discharge approximately 12 gallons per minute.

Spacing of Sprinklers.—For smooth ceilings, sprinkler outlets must be placed every ten feet each way of the building. This provides one head for each 100 sq. ft. of area. In paneled ceilings, one line of sprinklers must be placed in the centre of each bay, and the distances between the sprinklers on each line must not exceed the following, in which the widths of the bays are taken as centre to centre of th projecting beams or girders which form the panels:

> 12 ft. bays, 7 to 8 ft. centres;
> 11 ft. bays, 8 to 9 ft. centres;
> 6 to 10 ft. bays, 10 ft. centres.

Feed-mains and Risers.—There must be a separate riser for each building and for each section of a building divided by fire-walls. The size of each riser must be sufficient to supply all the sprinklers on any one floor at the same time. " Centre central " or " side central " feed to sprinklers is to be recommended. " End " feed is not approved by most of the insurance regulations.

The circulation of water in sprinkler-pipes is very objectionable, as greatly increased corrosion, deposit of sediment, and condensation-drip from the pipes result.

Water-supply for Sprinklers.—Acceptable water-supplies for sprinkler systems may be furnished by:

City water-supply;

Duplex steam-pump;

Private reservoir or stand-pipe;

Elevated gravity-tank;

Air-pressure tank, or,

Rotary pump.

Two independent water-supplies are absolutely essential for the best equipment. At least one of the supplies must be automatic and one must be capable of furnishing water under heavy pressure. The first, or automatic supply, is usually provided by the city pressure taken from the public water-mains, and the second, or pressure supply, usually consists of either an elevated gravity-tank or an air-pressure tank.

The public water-works supply should be sufficient to give a good pressure at all hours of the day to the highest line of sprinkler-heads, preferably not less than 25 pounds static pressure.

Gravity-tanks should be so placed that the bottom of the tank in each case should never be less than 12 ft. above the highest line of sprinklers which it supplies. Tanks of from 5,000 to 10,000 gallons capacity are usually employed. In plants of moderate size a tank of not less than 5,000 gallons is required, and in extensive plants, or where the secondary supply is limited in its capacity, either a larger tank or more tanks should be used.

Air-pressure tanks are located either in the top floors of the building or on the roof. The capacity should preferably be about 4,500 gallons. This requires a tank 72 ins. diam. and 22 ft. long, or 66 ins. diam. and 25 ft. long. The tank must be kept two-thirds full of water, and an air-pressure of not less than 75 lbs. must be maintained, so as to give not less than 15 lbs. pressure at the highest line of sprinklers when all water has been discharged from the tank.

Whatever the water-supply for sprinkler systems, sidewalk connections which permit of the direct attachment of the fire-engines to the risers, thus giving a constant supply under heavy pressure, are strongly to be recommended. A 10,000-gallon tank may, under severe circumstances, be emptied in ten minutes if 50 sprinklers happen to be open at once. Such a system, fed from a gravity or air-pressure tank, is worth hardly half as much as one fed by a 6-in. or 8-in. riser under a 75-lb. pressure.

Automatic Alarms.—Automatic sprinklers must be accompanied by either automatic alarms or by watchman service. This is to prevent the opening of the sprinkler-heads under the effects of heat from a small fire, and then continuing their constant downpour until discovered. No system of automatic shut-off has yet been invented. Cases have frequently occurred where a small fire has started during a Saturday night, in which the sprinkler soon extinguished the fire, but the water continued to run until discovered on the Monday morning following. This could easily occur where the sprinkler-head and the thermostat were in different bays. To avoid just such cases, either watchman service or automatic alarm must accompany the sprinkler system.

If a watchman's services are relied upon, record of trips must be made either by means of watch-clocks or by connection to a central registering office.

Automatic alarm should be provided by means of an alarm-valve, so constructed that the flow of water through but a single sprinkler would operate an electric gong or a mechanical gong, or both, as circumstances may require. Where there is a thermostat-alarm company with a central station, the alarm valve may be connected to such central station.

In connection with automatic sprinklers, it should be

needless to add that nothing should be placed upon or over the floor-space to prevent the proper distribution of the water from the sprinklers; and yet, during the holiday season a case was noticed in a large dry-goods store where ornamental booths had been so erected in the building for exhibition purposes that the sprinklers were efficiently shielded, so that fire could rapidly spread in the inflammable contents of the interiors before the sprinkler-heads could be opened for service.

Open Sprinklers.—Open sprinklers are principally valuable in case of exposure fire, or in preventing the passage of flames from floor to floor by means of the windows. The installation consists of vertical risers and horizontal feed-mains, placed within the building, with outlets on the exterior of the building, over the windows or other openings in the exterior walls.

The sprinklers consist of metal orifices placed over the openings, and just outside of the window-frames. Below the orifice, and forming a part of the sprinkler, is placed a horizontal shovel-shaped disk, the office of which is to deflect the water, as it is discharged, into a sheet or spray of sufficient area to protect the window below.

The system is free from water except when in use, at which times supply is had by opening a valve placed at a safe distance from exposure, to admit the pressure from the city mains, or from a fire-pump. Open sprinklers should never be connected with either pressure or gravity tanks used to supply automatic sprinklers. A sidewalk connection for a direct supply from steam fire-engine service would constitute a most commendable feature.

The risers should preferably feed the horizontal mains from the centre. The sprinkler-heads may be placed either in one line only, in which case they are run directly under the cornice, or in two lines, one at the cornice and one two or three stories below the cornice, or one head outside of and opposite every window on each floor. For buildings containing highly inflammable contents, or with dangerous exposure, the latter arrangement will be found desirable, and it is generally employed.

It is to be remembered that in a system of open sprinklers *all* are operated at one time, and with a system of any size there is a large quantity of water discharged per minute. Each open sprinkler must have an unobstructed outlet of such size and form that with 5 lbs. pressure maintained at the sprinkler it will discharge approximately 6 gallons per minute. The sprinklers should not be located over 8 ft. apart on the line, as the spray will not exceed this width.

Dry-pipe Systems. — Where buildings or portions of buildings are so constructed, or where the nature of the occupancy is such that the premises cannot be kept sufficiently warm to prevent the water in sprinkler-pipes from freezing, the ordinary wet-pipe system cannot be used. A valve, called a dry-valve, is then introduced to keep the water out of the pipes. The sprinkler-pipes are filled with air under pressure, and the reduction of this air-pressure, through the opening of a sprinkler-head, allows the water pressure to open the dry valve automatically, thus admitting water to the sprinkler system to be distributed through open sprinklers.

A dry-pipe system is not to be recommended where a wet-pipe system can be used.

Stand-pipes and Hose-reels.—Automatic-sprinkler systems are particularly applicable to manufacturing and mercantile buildings, and their use is to be strongly recommended in this class of structures; but in hotels, office buildings, etc., where sprinkler systems are hardly suitable, other means of adequate protection must be devised.

The difficulty experienced in fighting fire in tall buildings may be said to increase about as the square of the height.

Fire apparatus in present use is wholly inefficient at elevations above 125 ft. from the street level, and yet almost all of our larger cities contain structures which rise to considerably greater heights. Means must therefore be obtained to secure suitable protection in the upper floors of very high structures, and this requires that equipment must be provided toward this end.

This equipment for fireproof buildings which are not supplied with automatic sprinklers, can best be obtained through the use of stand-pipes with hose-connections or hose-reels on each and every floor. These devices become absolutely necessary to fight fire above about the tenth story, and even for the lower floors they prevent the necessity of the firemen having to drag their hose up tortuous and inconveniently placed stairways, and the attendant difficulty and exhaustion experienced, as in the Vanderbilt Building fire, or the dependence for the elevation of their fire-hose upon elevators which are not in service at the times of such crises.

The lesson taught by the Home Life Insurance Building fire and others in high buildings—that above an elevation of about 125 ft. the fire must be left to spread at random or burn itself out—shows the importance of furnishing efficient means of extinguishing fire at any and all floors through the provision of a sufficient water-supply under adequate pressure. In very many cases, interior apparatus and a stand-pipe system have been installed, but the service rendered is usually inefficient because the size and capacity of the system is not equal to the magnitude of the work to be accomplished. Again, it is doubtful in how many cases of high buildings sufficient water-pressure is available to supply the requisite power at the nozzle for even a few of the upper stories in a building from 20 to 30 stories in height. Efficient service would necessitate a separate fire-pump for such emergencies, which the owner would no doubt consider a disproportionately burdensome safeguard; so the water-supply pump is relied upon to do doubtful duty for a fire-pump as well, and if the test is a severe one, the supply of water is utterly insufficient. It has been suggested * that where hydraulic elevator service exists, the elevator-pumps be so arranged that they can be thrown onto the fire service in case of need. But in many cases both the building and the elevator-pumps are liable to become inoperative through the flooding of the basement, or through accident, and this possibility shows the supreme importance of being able to connect the stand-pipe system directly to the fire-department engines.

Fire Marshal Swenie of Chicago, in urging the necessity of a stand-pipe system, recommends that the stand-pipes be not less than 6 ins. in diameter, that they should have Siamese connections at the street-level for the fire department, that they should be provided with check-valves, so that when the department engines are attached their force will be added to that of the building's pumps; that there should be hydrants on each floor and on the roof, and that each floor should have hose-connections with the stand-pipe and sufficient hose to reach the most remote point of the floor above or below.

For use in exposure fires, Monitor nozzles placed on the roof and connected to the stand-pipe system form a valuable means of protecting the exposed faces of the building. These nozzles are fixed in position, but may be raised or lowered by means of a lever, or revolved horizontally for use in any direction.

Ex-Chief Bonner, of the New York Fire Department, has made some recent tests of the stand-pipe system of the Manhattan Building. Tests were made with the pumping-engines of the building itself and with a fire-department engine.

* See " The Need of Improved Methods for Extinguishing Fire in High Buildings," *Engineering News*, Dec. 22, 1898.

With the latter power, water was forced through the stand-pipe to the dome gallery, 340 ft. above the street-level, and from this elevation a stream was thrown about 70 ft. above the top of the flagstaff, or to a height of 450 ft. above the street-level. The pressure at the 1⅝-in. diam. nozzle was 50 lbs.

In the Home Life Building fire it was found that when fire-hose was carried to the upper floors and the attempt was made to force water to such a height, the hose burst, while in the tests made by Chief Bonner, the stand-pipe stood the test most satisfactorily. If efficient stand-pipe systems were installed in all high buildings, they would add greatly to the safety from fire. Their use, as well as the installation of salt-water pipe-lines for the purpose of furnishing an adequate water-supply has been recommended by the Board of Fire Underwriters in New York.

Auxiliary-pipe Systems.—Although not distinctly a part of equipment as applied to fireproof buildings, auxiliary-pipe systems demand careful consideration inasmuch as such systems render much more efficient the stand-pipes installed in high structures, and furnish a much higher pressure for the ordinary fire-hose handled by the fire departments.

In a recent report of the New York Fire Department, Chief Bonner speaks as follows regarding auxiliary-pipe lines:

" There are high buildings in New York that are entirely beyond the ability of the machinery of the fire department to force water to the upper stories. It must also be considered that the tall buildings of the present day, as well as the mercantile buildings, are of greater height and cover larger areas than ever before, and that fires will entail great loss if not promptly checked. If the department is not able to control the fire and surround the structures with large quantities of water, there is a possibility that such fires will extend and become great conflagrations. This source of danger can be obviated by adding an auxiliary plant for the use of the department in the form of the underground-pipe line, connected to a roof-line system, as well as an aerial line, which should connect to all of the tall buildings. This should be fed from the underground system and also from our city mains, with the aid of the auxiliary pumps—electric or steam—located at different points in the city. By the use of the auxiliary pump, water could then be forced to the topmost story of any building, no matter how high, as well as to the circuit on the roof-line, and would render positive protection against any extensive fire in that section of the city.''

Auxiliary-pipe systems have already been installed in Boston, Detroit, Cleveland, Milwaukee, and Buffalo, with apparently satisfactory results. Where these lines may be fed by an inexhaustible supply of fresh water, as is the case in all of the cities above named except Boston, this auxiliary service may be connected directly with or made a part of the regular city supply, and the additional pressure, obtained by means of high-duty pumping-stations or through the use of powerful fire-boats, may be utilized by the fire department both in their own hose and in the stand-pipe systems. But where a salt-water pipe-line is employed, as is the case in Boston, this supply cannot be utilized for stand-pipes, as the corrosive influences of the salt water and the rusting of the valves and fittings would probably preclude the use for such purposes. The Boston pipe-line is used as an auxiliary supply for the department hose only, and is filled only as occasion demands.

Very recent tests of the system in Boston have demonstrated that with this salt-water supply, fires can be successfully fought at a much longer distance from the hydrants than with the fresh-water steamers, thereby avoiding the massing of so many fire-engines close to the fire, as is now the custom. One test consisted of attaching two lines of 3-in. hose, each

800 ft. long, to a Siamese 1½-in. nozzle; and to the same hydrant one line of 2½-in. hose only 200 ft. long was attached. The fire-boat, which furnished the pressure, was fully one mile away from the nozzles, measured on the underground main and hose. The pressure at the fire-boat was 200 lbs., at the main 188 lbs., at the hydrant outlet 175 lbs., at the Siamese 90 lbs., at the nozzle of the 800-ft. line 85 lbs., and at the nozzle of the 200-ft. line 83 lbs.

Accessibility.—In Chapter VII it was recommended that stair-wells be isolated in all possible cases, and that they be conveniently arranged, so as to be easily accessible and easily mounted. This is for the benefit of the fire department, as well as for the protection of the inmates. But for buildings of great height, stairways will not suffice for the proper working of the fire department. The difficulty of ascending many flights of stairs, even without the added burden of hose, may, under considerations of smoke or exertion, seriously embarrass the prompt working of the department.

In all buildings of considerable height elevator service should be available at all hours. This means that sufficient power must be had at all times for the operation of at least one elevator, and that employees, competent to furnish this service, and to aid and direct the firemen in case of fire, be constantly on hand. The one elevator which was used at the time of the Home Insurance Building fire proved of great value until the fire in the upper stories rendered its use impossible. If the elevators as well as the stairways were enclosed in isolated shafts, and protected by fire-doors, the disabling of the elevator service could be avoided, and an added protection would be furnished through the accessibility provided.

Inspection.—Automatic fire protections and stand-pipes,

hose-reels, etc., require a careful inspection at suitable intervals to preclude the possibility of failure in time of emergency.

The failures in sprinkler systems are most common where
exposed to corrosive vapors such as occur in glucose factories
or sugar refineries. A periodical inspection of sprinklers is
always necessary. In retail or department stores, tests cannot
be made by opening the heads, and in these cases it is necessary that the installation should be under the constant
supervision of one thoroughly acquainted with their construction. Any trouble in the head may be discovered by
inspection.

The Boston Board of Fire Underwriters inspects all
department stores once in four weeks, and other risks once in
six weeks. A head is considered in good condition if it will
operate at a temperature starting at 70 degrees F. and increasing to 250 degrees in two minutes.

Automatic fire-alarm systems are always regularly inspected and tested by the fire-alarm companies.

As regards stand-pipe service, and hose-reels, monitor
nozzles, etc., a proper inspection of these devices by the fire-
department authorities is always necessary, and it has been
suggested that the control and supervision of these details of
equipment be placed entirely within the jurisdiction of the
fire department in the event of fire.

Discipline of Employees.—Thorough acquaintance with
all methods of fire protection and details of fire equipment,
etc., should be required of the employees who care for the
building. Frequent practice or drill in the use of the apparatus at hand may be of great value in caring for incipient
fires, or in aiding the fire department to locate or operate
equipment with which they are not thoroughly familiar.

INDEX.

	PAGE
Anchorage of walls..	260
Arches, brick...	9
composition—see Composition arches.	
concrete—see Concrete arches.	
corrugated iron..	10
segmental terra-cotta—see Segmental terra-cotta arches.	
terra-cotta—see Terra-cotta arches.	
Asbestic plaster, composition of........	101
fire-resistance of	102
Attic spaces..	283
Automatic alarms...	300
fire-alarm systems	293, 308
sprinklers...	296
Auxiliary pipe-systems...	305
Bailey floor, tests of...	66
Balconies...	267
Bay-windows..	267
Brick arches..........	9
fire-resistance of...	83
in exterior walls...	259
permanency of different grades of............................	115
British Fire-prevention Committee.................................	6
Cable Building, columns in exterior walls.........................	270
lintel protection.....................................	265
spandrel sections in.......................... 264, 265,	266
Cage construction—see Skeleton construction.	
Camber in terra-cotta arches.....................................	165
Cast-iron columns in exterior walls................................	256
tests of..	78
corrosion of..........	110
fire-resistance of...	77
Ceilings combined with roofs......................................	274
metal furring for......................................	289
suspended..	284
from roofs..	277
terra-cotta blocks used for...............................	287
Cement-mortar, corrosive action of................................	112
fire-resistance of............................. 96, 100,	104

PAGE

Centering for Columbian arches.. 194
          Expanded Metal Co.'s floor.................................. 187
          Metropolitan floor... 197
          Roebling floor... 184
          terra-cotta arches.................................... 164, 167
Chicago Athletic Club Building, damage to exterior walls............... 259
                         fire in............................... 28, 138
Cinder-concrete.. 119
         partitions.. 239
Cinders.. 106
Columbian floor, advantages and disadvantages in.............. 195, 211
           cinders vs. stone....................................... 106
           description of.. 190
           method of setting. 194
           tests of 65
           weights of... 194
       method of column protection. 221
Column-guards... 234
Columns, fireproofing of, composition casings for..................... 224
           concrete protection for................................. 221
           development of... 17
           essentials for... 218
           hard vs. porous terra-cotta in......................... 231
           importance of... 217
           metal lath and plaster for............................. 219
           pipe spaces in... 232
           solid vs. hollow terra-cotta in........................ 231
           terra-cotta casings for................................. 224
       in exterior walls.. 267
       protection of interior..................................... 223
       tests of... 78
Comparison of concrete and composition floors......................... 208
Composition arches, choice of... 208
           Guastavino floor. 206
           Hecla floor.. 206
           Lee hollow-tile and cable-rod floor.................... 206
           Mackolite floor—see Mackolite floor.
           Metropolitan floor—see Metropolitan floor.
           Poulson floor.. 206
Concrete arches, choice of.. 208
           classes of... 178
           Columbian Fireproofing Co.'s floor—see Columbian floor.
           Expanded Metal Co.'s floors—see Expanded Metal Co.'s
             floors.
           Fawcett system—see Fawcett floor.
           Melan arch—see Melan arch.
           Rapp floor—see Rapp floor.

PAGE

Concrete arches, Roebling arch—see Roebling arch.
    flat construction............................... 201
    tests of.................................... 51, 56
    vs. terra-cotta arches................................. 212
    without reinforcing members.......................... 179
  column protection.................................. 221
    Columbian method......................... 221
    Roebling method..... 221
  construction 22
  corrosive action of....... 118
  fire-resistance of............................ 102
  partitions................................. 239
  roof construction............................. 275
  vs. terra-cotta 212
Cornice furring................................ 289
Corrosion, causes of................................ 110
  of cast- and wrought-iron and steel...................... 110
  protection against................................ 108
  relation to fireproofing.......................... 109
Corrugated-iron arches................................ 10
  shutters................................. 132
  doors................................ 130
Counters in stores................................ 123

Door-frames for fireproof partitions....................... 252
Doors, corrugated-iron................................ 130
  fireproof................................ 129
  in fireproof partitions........................ 251
  in stair- and elevator-wells...................... 139
  in wall-openings........................... 261
  metal-covered............................ 130
  sliding................................ 130
  tin-covered............................. 129
Druecker warehouses, column protection in.................. 222
Dry-pipe sprinklers............................ 302

Elevator gratings................................ 141
Elevators, efficient service for fire..................... 307
Elevator-shafts, design of......................... 137
  doors in 139
  ironwork in............................ 257
Ellicott Square Building, protection of column interiors............ 224
Equipment for fire-resistance........................ 292
Expanded Metal Co.'s floors, advantages and disadvantages of...... 188, 211
  description of............................ 185
  tests of............................. 68
  partitions, hollow-plaster........................ 241

PAGE

Expanded Metal Co.'s partitions, solid-plaster.......................... 240
       roofs.. 275

Fair, The, building, column fireproofing in......................... 229
Fawcett floor, advantages and disadvantages of...................... 205, 210
     description of.. 203
     tests of.. 71
Feed-mains and risers... 298
Filling blocks.. 158
Fire-alarm systems, automatic....................................... 293, 308
Fire insurance—see Insurance.
Fire losses, annual in United States................................. 1
Fireproof buildings, earliest types of.............................. 8
     origin of present methods.... 9
Fireproofing, imperfect methods of... 112
   materials used in—see Materials.
Fire-resistance—see Materials.
     equipment for.. 292
Fires, Chicago Athletic Club Building............................. 28, 138
   Home Life Insurance Building, New York.... 42, 85, 100, 136, 163, 171
   Hotel Windsor.. 2
   in Europe and Great Britain.................................. 3
   Livingston Building, New York..... .... 40, 100, 138
   Lumber Exchange Building, Minneapolis....................... 26
   Manhattan Savings Bank Building............................ 24, 142
   Pittsburg buildings 31, 78, 80, 85, 127, 128, 134, 140
   Schiller Theatre Building, Chicago......................... 28, 92
   Vanderbilt Building..... .... 41, 137, 142
Fire walls.. 127
Fisher Building, columns in exterior walls......................... 270
     roof and cornice construction....................... 284
Flange protection, earliest type of................................. 13
     side-construction arches............................ 150
Floors—see Arches.
     comparative weights of different types..... 207
Furring... 289
Furring-blocks.. 261

Girder protection, Columbian system................................. 191
     importance of.. 172
     of plate- and box-girders........................... 175
     terra-cotta.. 173
Glass, wire... 124
Granite, fire-resistance of... 82
     in exterior walls................................... 258
Great Northern Hotel and Theatre building, bay-windows in........... 267
Guastavino floor, tests of.. 72
     use of.. 206

PAGE

Havemeyer Building, columns in exterior walls...................... 269
Hecla floor..... 206
Home Life Insurance Building, brick walls in....................... 85
    concrete floor filling in.............. 163, 171
    damage to exterior walls................ 259
    exterior walls........................... 257
    fire in.................................. 42
    partitions in....................... 42, 100
Hose-reels... 302, 308
Hotel Windsor, fire in... 2

I-beams, invention of... 9
    first rolled in United States............... 9
Insurance, annual amount of paid............................... 1
    rates of in Boston......................... 4
Iron in exterior walls... 256

Lee hollow-tile and cable-rod floor............................ 206
Light-shafts, interior... 127
    ironwork in exterior....................... 257
Lime-mortar, corrosive action of............................... 113
    fire-resistance of......................... 100
Limestone, corrosive action of................................. 114
    fire-resistance of......................... 82
    in exterior walls.......................... 258
Lintels, protection of... 265
Livingston Building, column protection in...................... 220
    fire in.................................... 40
    plaster partitions in.................. 100, 138
Lumber Exchange Building, fire in.............................. 26

Mackolite column protection.................................... 224
    floor, description of...................... 189
    weight of.................................. 190
    partitions................................. 253
Manhattan Building, stand-pipe tests in........................ 304
    floor, tests of............................ 65
    Savings Bank Building, fire in.......... 24, 142
Marble, fire-resistance of..................................... 82
    in exterior walls.......................... 258
    stair-treads............................... 142
Masonry—see Brick, Granite, Limestone, Marble, Stone.
Massachusetts State House, furring in.......................... 290
Materials for fireproofing, brick.......................... 83, 122
    cast-iron.................................. 77
    concrete............................... 102, 122
    fireproof wood............................. 123
    in exterior walls.......................... 256

PAGE

Materials for fireproofing, mortars.. 96, 99
      ornamental terra-cotta........................... 92
      plaster of Paris 96, 100
      plasters....................................... 96, 99
      porous vs. hard-burned terra-cotta............. 91
      steel.. 77
      stone.......................................,,.. 81
      terra-cotta................................ 85, 122
      wire-glass..................................... 124
      wrought-iron 77
M'Cabe floor, tests of........ 64
Mechanical plants, install'ation of..................................... 143
Melan arch, description of.. 180
      tests of................................... 56
      with metal lath.................................. 181
Metropolitan floor, advantages and disadvantages of................. 198, 210
      description of.................................... 196
      method of setting................................ 197
      tests of............................ .... 53, 70
      weight of.. 198
      partitions...................................... 253
Mill construction....................................... ... 21
      tests of floors................................. 75
Monadnock Building, column fireproofing in.......................... 233
Monitor nozzles.. 304, 308
Mortars, corrosive action of........................... 112
      fire-resistance of.................................. 96

Nailing-strips... 162
New York Building Department tests.......... 58
New York Life Building, Chicago, column fireproofing in............... 227

Pabst Building, ironwork in... 258
Partitions around stair- and elevator-shafts 137, 139
      composition..................................... 253
      hollow-plaster................................... 241
      plaster, Expanded Metal system................. 240, 241
        Roebling................................ 238, 241
      requirements for................................ 236
      solid-plaster.................................... 238
      sound-proof..................................... 252
      terra-cotta...................................... 224
        essentials for.............................. 250
        method of setting.......................... 247
      tests of in Home Life Building.................... 100
Party-walls... 260
Pent-houses.. 288

PAGE

Pipe-shafts.. 143
    spaces in column fireproofing................................. 232
Piping in column casings.. 232
    exterior walls... 272
    installation of.. 143
    moisture from.. 120
Pittsburg buildings, brickwork in.................................... 85
    cast-iron columns in.. 78
    distortion of framework in,................................. 80
    fire in... 31
    fire-shutters in.. 133
    light-shafts in... 127
    protection of spandrels in.................................. 267
    relative positions of....................................... 32
    report on terra-cotta vs. concrete.......................... 212
    roof construction in.. 279
Plaster column protection... 219
    fire-resistance of... 96
    partitions, hollow.. 241
        in Home Life Building............................. 100
        solid... 238
        value of.. 242
Plaster of Paris column protection.................................. 224
    fire-resistance of................................... 96, 100
    partitions.. 253
    used in Metropolitan floor................................... 196
Poulson floor... 206

Rapp floor, advantages and disadvantages of.................. 201, 209
    description of.. 199
    test of... 61
    weight of... 201
Reliance Building, column fireproofing in........................... 233
Roebling arch, advantages and disadvantages of.............. 184, 212
    description of.. 181
    tests of.. 61
    weight of... 184
    column protection... 221
    flat construction, description of............................. 201
    partitions, hollow.. 241
        solid... 238
Roofs and ceiling combined.. 274
    attic spaces.. 283
    construction of... 273
    coverings for... 288
    in U. S. Appraisers' Warehouse................................ 278
        Public Buildings, Portland, Ore................... 275

PAGE

Roofs in U. S. Public Building, San Francisco.......................... 275

  mansard.. 283

  pent-houses for.. 288

  pitched.. 279

  suspended ceilings... 284

  terra-cotta blocks used in 287

  with suspended ceilings...................................... 277

Roof trusses, fireproofing of.................................... 280

Schiller Theatre Building, fire in...........................28, 92

Scoring of terra-cotta arches 150, 162

Segmental terra-cotta arches, construction of.................... 159

  early forms of............................ 15

  tests of..53, 54

  weights of................................ 160

  with flat ceilings 161

Shaft-openings.. 261

Shelving in stores.. 123

Shutters, asbestos.. 128

  corrugated-iron 132

  fireproof 132

  in wall-openings 261

  objections to.. 136

  rolling steel...128, 134

  tin-covered.. 132

  wood.. 128

Skeleton construction, definition of............................ 18

  development of................................ 20

  origin of..................................... 19

Skew-backs, raised.. 158

Spandrels... 263

Sprinklers, automatic......................................296, 308

  dry-pipe 302

  open.. 301

Stairs, design of... 141

Stair-wells, accessibility of................................... 307

  design of... 137

  doors in.. 139

  ironwork in 257

Stand-pipes..302, 308

  in elevator-wells.. 141

Steam-piping 145

  in exterior walls... 272

Steel, annual tonnage of, in United States...................... 108

  corrosion of.. 110

  expansion of.. 80

  fire-resistance of 77

PAGE

Stone, fire-resistance of.. 81
    in exterior walls.. 258
    permeability, porosity, and chemical action of.................... 114
St. Paul Building, columns in exterior walls........................... 269
Subdivision of large areas.. 125
Suspended ceilings...277, 284

Terra-cotta arches, camber of... 165
    ceiling finish... 162
    combination construction 153
    comparison in methods of manufacture................ 156
    defects in.. 167
    early types of...11-16
    end-construction method 150, 157
    flat construction of.. 148
    floor finish... 162
    inspection required... 168
    introduction of... 10
    method of setting............ 164
    segmental—see Segmental terra-cotta arches.
    selection of type....................................... 169
    side-construction method 149, 156
    strength of side- *vs.* end-construction.............157, 170
    tests of—see Tests.
    vs. concrete arches................................. 212
    weather and stain protection......................... 164
  ceiling tile........ 287
  column-casings, cast-iron columns, circular finish.......... ... 224
                rectangular finish.......... 227
    solid *vs.* hollow.......................... 231
    special shapes............................. 230
    steel columns, circular finish................. 226
          rectangular finish............... 229
  floors—see Terra-cotta arches.
  furring-blocks... 261
  hard-burned, characteristics of............... 147
    method of manufacture....... 90
    vs. porous..................................... 91
    for column-casings.................. 231
  in exterior walls... 259
  ornamental............. 92
    fire-resistance of................................. 95
    method of manufacture........................ 92
  partitions, description of.................................. 244
    essentials for.................................. 250
    method of setting.............................. 247
  permeability, porosity, and chemical action of................ 115

PAGE

Terra-cotta porous, characteristics of............................... 147
          introduction of...................................... 15
          method of manufacture............................... 85
          vs. hard-burned....   91
                    for column-casings.................... 231
     roofing tile...276, 287
     varieties of... 85
Test-kilns... 59
Tests, Denver...49, 91, 157
     general requirements for...................................... 49
     Geo. Hill 56, 170
     New York Building Department.............................. 58
     Tests of auxiliary pipe lines................................ 306
               Bailey floor..................................... 66
               building stones and terra-cotta...................... 82
               cement and cement-mortar........................ 96
               Columbian floor................................. 65
               concrete arches................................51, 56
                    floors (Hamburg).......................103, 119
               Expanded Metal Co.'s floor...................... 68
               Fawcett floor 71
               Guastavino floor................................ 72
               Manhattan floor.................................. 67
               McCabe floor.................................... 64
               Melan floor..................................... 56
               Metropolitan floor.............................53, 70
               Rapp floor...................................... 61
               Roebling floor.................................. 61
               segmental terra-cotta arches..................53-54
               stand-pipes.....   304
               terra-cotta arches..................49, 54, 56, 71, 72, 73
               Thomson floor 63
               wooden floor construction... 74
     St. Louis... 51
     Trenton 53
     Von Emperger.. 54
Thermostats 293
Thomson floor, test of.. 63
Tie-rods for terra-cotta arches................................... 161
Trusses, fireproofing of.. 280

United States Appraisers' Warehouse, column fireproofing in........... 223
                              rolling shutters in................. 135
                              roof construction of.............. 278
     Public Building, Portland, Ore., plate-girder protection in... 176
                              roof construction in....... 275
          San Francisco, concrete floors in........ . 187

PAGE

United States Public Building, San Francisco, partitions in.............. 251

roof construction in........ 275

Vanderbilt Building, fire in... 41

fireproof shutters in............................. 137

stairways in.................................... 142

Walls, anchorage of... 260

columns in..... 267

exterior... 255

materials for.. 256

openings in... 261

party... 260

spandrel.. 263

thickness of.. 260

Water-supply for sprinklers... 299

Weights, comparative, of different floor types........................ 207

of terra-cotta arches..........................150, 152, 153, 155

Window-frames, fireproof... 137

shutters—see Shutters.

Wire glass...124, 137

Wooden floors, methods of laying..................................... 163

tests of.... 74

Wood, fireproof... 123

Wrought-iron, corrosion of.. 110

fire-resistance of... 77

www.ingramcontent.com/pod-product-compliance
Lightning Source LLC
Chambersburg PA
CBHW021501210326
41599CB00012B/1093